AA002337

2020 IEEE 3rd Conference on PhD Research in Microelectronics and Electronics in Latin America (PRIME-LA 2020)

San Jose, Costa Rica
25-28 February 2020

IEEE Catalog Number: CFP20K02-POD
ISBN: 978-1-7281-3147-4

**Copyright © 2020 by the Institute of Electrical and Electronics Engineers, Inc.
All Rights Reserved**

Copyright and Reprint Permissions: Abstracting is permitted with credit to the source. Libraries are permitted to photocopy beyond the limit of U.S. copyright law for private use of patrons those articles in this volume that carry a code at the bottom of the first page, provided the per-copy fee indicated in the code is paid through Copyright Clearance Center, 222 Rosewood Drive, Danvers, MA 01923.

For other copying, reprint or republication permission, write to IEEE Copyrights Manager, IEEE Service Center, 445 Hoes Lane, Piscataway, NJ 08854. All rights reserved.

*** This is a print representation of what appears in the IEEE Digital Library. Some format issues inherent in the e-media version may also appear in this print version.

IEEE Catalog Number: CFP20K02-POD
ISBN (Print-On-Demand):   978-1-7281-3147-4
ISBN (Online):   978-1-7281-3146-7

Additional Copies of This Publication Are Available From:

Curran Associates, Inc
57 Morehouse Lane
Red Hook, NY 12571 USA
Phone: (845) 758-0400
Fax: (845) 758-2633
E-mail: curran@proceedings.com
Web: www.proceedings.com

2020 IEEE 3rd Conference on PhD Research in Microelectronics and Electronics in Latin America (PRIME-LA 2020)

San Jose, Costa Rica
25-28 February 2020

IEEE Catalog Number: CFP20K02-POD
ISBN: 978-1-7281-3147-4

PRIME-LA 2020

3rd IEEE Conference on Ph.D. Research in Microelectronics and Electronics in Latin America

Holiday-Inn Hotel, Plaza Tempo, Escazú
San José, Costa Rica
February 25-28, 2020

Conference Proceedings

http://www.ie.tec.ac.cr/lascas2020

Table of Contents

PRIME-1	Carlos Salazar-García, Jefferson González-Gómez, Kaleb Alfaro-Badilla, Ronny García-Ramírez, Renato Rimolo-Donadio, Christos Strydis, and Alfonso Chacon-Rodriguez. **PlasticNet: A Low Latency Flexible Network Architecture for Interconnected Multi-FPGA Systems.**	1
PRIME-2	Maximiliano Chiossi and Matías Miguez. **Low-Power Activity Recognition from Triaxial Accelerometer Data.**	5
PRIME-3	William Teles Medeiros, Hamilton Klimach, and Sergio Bampi. **Ultra-Low Power Relaxation Oscillation Survey: Design Trends and Challenges.**	9
PRIME-4	Ronny García-Ramírez, Alfonso Chacon-Rodriguez, Christos Strydis, and Renato Rimolo-Donadio. **Pre-Synthesis Evaluation of Digital Bus Micro-Architectures.**	13
PRIME-5	Roberto Molina-Robles, Edgar Solera-Bolaños, Ronny García-Ramírez, Alfonso Chacon-Rodriguez, Alfredo Arnaud, and Renato Rimolo-Donadio. **A Compact Functional Verification Flow for a RISC-V 32I Based Core.**	17

PlasticNet: A low latency flexible network architecture for interconnected multi-FPGA systems

Carlos Salazar-García*, Jeferson González-Gómez[†], Kaleb Alfaro-Badilla[‡],
Ronny García-Ramírez[‡], Renato Rímolo-Donadío[‡], Christos Strydis[§], and Alfonso Chacón-Rodríguez[‡]

*Mechatronics, [†] Computer, [‡] Electronics Engineering, Instituto Tecnológico de Costa Rica, Cartago, Costa Rica
[§]Dept. of Neuroscience, Erasmus Medical Center, Rotterdam, The Netherlands
Email: {csalazar, jgonzalez, jkalfaro, rgarcia, rrimolo, alchacon}@tec.ac.cr, {c.strydis}@erasmusmc.nl

Abstract—This paper presents preliminary results of Plastic-Net, a custom FPGA interconnect architecture designed for high-processing applications that communicate extensively among multiple FPGAs. PlasticNet allows the interconnection of processing nodes (PNs) through a flexible, reliable and efficient custom protocol, that can be easily integrated in High-Level Synthesis (HLS) modern design environments. The system is evaluated on a ZedBoard Zynq®-7000 ARM/FPGA SoC Development Board, including criteria such as overhead, area, worst-case packet delivery latency and bandwidth. The best evaluated case achieved a half-occupancy latency of $16.9 \mu s$. The results show the potential of PlasticNet as an efficient solution for low latency multi-FPGA interconnection.

Index Terms—Customized network protocol, HLS, interconnect architecture, inter-FPGA communication, multi-FPGA.

I. INTRODUCTION

Despite modern FPGA's high-logic capacity, floating-point capabilities, integration of standard processing cores and significant improvement of development tools such as High-Level Synthesis (HLS) for the rapid prototyping of complex designs (see [1]–[4]), single FPGAs are not enough when processing becomes massive. For this reason, digital design across multi-FPGA systems becomes mandatory in massive applications implying high communication requirements.

This paper presents preliminary results of PlasticNet, a custom FPGA interconnect architecture aimed at applications that, by their nature, may be partitioned into different PNs that communicate extensively with each other within the same FPGA or among multiple FPGAs. PlasticNet provides the communication resources between the different PNs that make up the system, with reliable transmission of messages across FPGA boundaries, taking advantage of their interconnection resources. Additionally, PlasticNet can be seamlessly integrated into HLS design environments, being fully compatible with most external interfaces supported by HLS tools.

This paper is organized as follows: Section II compares multi-FPGA systems reported in the literature. Section III reviews of a variety of methods for physical inter-FPGA communication found in the state-of-the-art. Section IV details the communication requirements of PlasticNet. Section V shows an overview of the proposed implementation. In Section VI, PlasticNet's performance is tested for different configurations.

Lastly, Section VII highlights the main conclusions and discusses future works.

II. COMPARISON OF DIFFERENT MULTI-FPGA APPROACHES

According to [5], multi-FPGA systems may be classified from the point of view of their interconnections in three different groups: Hardwired Off-the-Shelf, Custom and Cabling.

The first group is made up of a ready-made multi-FPGA board where each FPGA is connected with each other by fixed printed hardwires. Such solutions provide adequate performance, substantially accelerating development. However, they tend to be expensive and often require proprietary tools, making hard design migration among platforms. Besides, being connections among different FPGAs fixed, it is impossible to customize their interconnection architecture [6] to extend their functionality.

On the other hand, a custom multi-FPGA platform is defined as a build-your-own multi-FPGA board technique, where all the FPGAs are connected by printed hardwires. In this category, interconnections are user-defined and customized for a specific design. These platforms are far more complex to develop, require higher development times (increasing total time-to-market [7]) and designers with large experience in multilayer PCB design.

Finally, a cabling multi-FPGA platform is defined as the FPGA interconnection of several ready-made boards by using external cables. In this category, each board has a single FPGA and can be interconnected to other boards through wires, by external connectors. These ready-made single boards (also known as evaluation boards) are sold for rapid prototyping but they have become more popular for research and development purposes [8].

Table I compares the different categories mentioned above. After analyzing this information, we chose for this proposal to implement a multi-FPGA system using several ready-made single FPGA boards connected by external cables. Although the performance of this type of systems is not the highest, this is compensated by its price and development time.

III. PHYSICAL INTER-FPGA COMMUNICATION

The easiest option for communication in multi-FPGA stacks is using standard protocols, e.g., USB, Ethernet and PCI

978-1-7281-3147-4/20 $31.00 © 2020 IEEE

Table I
EVALUATION OF DIFFERENT MULTI-FPGA SYSTEM CATEGORIES.

Multi-FPGA Category	Scala-bility	Unit Price	Perfor-mance	Deploym-ent Cost	Reconfi-gurability
Custom	Low	Medium	High	high experience design team	Low
Hardwired Off-the-Shelf	Medium	High	Medium	high but it do not depend of the user	Low
Cabling	High	Low	Low	Low	High

Express. Evaluation boards usually incorporate physical interconnection ports for these protocols. For instance, authors in [9] developed a multi-FPGA system where communication between the PC and the FPGA master was implemented by means of a USB interface. Ethernet is a common option too, used for communication with a host PC or for inter-FPGA communication without high bandwidth requirements. Authors in [10] implemented a multi-FPGA system where inter-FGPA communication was carried out using Ethernet, with a maximum performance of 1Gbps. Lastly, PCI Express (PCIe), is perhaps the most used protocol for communication between an FPGA and a host PC. However, its implementation and use directly from the hardware is non-trivial. Additionally, most FPGA boards have a single PCIe port, which limits the communication among different boards. One example of these systems can be found in [11].

Since FPGA-based evaluation boards already provide different physical interfaces to achieve communication between different modules (ranging from Low Voltage Differential Signaling (LVDS) to high-speed links such as FPGA Mezzanine Card (FMC) connectors), we selected this option as the physical connection scheme between the boards, rather than using a more complex and expensive alternative. A limitation of this approach is the physical cabling distances between nodes, but to compensate for this issue, we have chosen a ring topology, which requires short interconnection paths between the boards. Using a simpler protocol also allows us to reduce the area utilization, increasing the overall number of PNs within the chip.

IV. COMMUNICATION REQUIREMENTS

Based on the above considerations, the main requirements for PlasticNet are:

1) A user-defined packet size: With the idea of building data-processing applications, in which hardware implementation may be partitioned into multiple PNs distributed among several FPGAs, packet data size must be completely reconfigurable according to each need.
2) Hardware-only: In order to leverage the high-logic capacity of modern FPGAs, all message transport reliably is to be carried out without resorting to a software layer.
3) Efficiency: PlasticNet must take full advantage of the variety of communication resources included in most of modern evaluation boards.
4) Reliability: Although the probability of an error is extremely low with a direct, short cabling interface, PlasticNet must properly handle those situations.

5) Reconfigurability: PlasticNet must be easily reconfigurable depending on the needs of each application. Additionally, PlasticNet must support the implementation of the different PNs using HLS tools.
6) Interoperability: PlasticNet must be compatible with different FPGA boards.

V. OVERVIEW OF THE HARDWARE DESIGN

Figure 1 provides a glimpse of the proposed PlasticNet architecture. Specifically, Fig. 1(a) shows the hardware blocks used to build PlasticNet's communication infrastructure, while Fig. 1(b) outlines its protocol hierarchy. User-defined applications are located in the highest layer, which in this paper's case, constitute a set of hardware-software PNs, each of them implementing a part of the process to accelerate. Each PN may be simply a module implemented in some hardware description language (HDL), or a complex block implemented using synthesized high level languages, integrated into heterogeneous SoCs. PlasticNet provides the flexible required interfaces for each approach, as either HDL libraries or synthesized HLS IPs that can be managed by Xilinx's Vivado design suite, with the required software drivers for its integration to a SoC-based PN.

The transport layer is in charge of the communication within each FPGA. This includes the unpacking/packaging unit, routing tables for each PN, and one internal bus which brings data from one PN to another. Within this layer, each PN sends messages of up to 8kB, where each message will be subsequently divided into flits of 128 up to 8192 bits depending on each application, thus maximizing performance without sacrificing reliability. In addition to the data to be transmitted, each PN will place two identifiers in the flit: TX_UID stands for the global identifier of the transmitter PN and the RX_UID stands for the global identifier of the receiving PN. Both TX_UID and RX_UID are unique for the entire multi-FPGA system and have a maximum size of up to 8 bits statically defined for each application. Each PN is connected via FIFO with the message unpacking/packaging unit. The main function of this block is either to split the message to be transmitted by the PN in different flits or to join the flits received from the internal bus to create a message. An extra identifier of up to 8 bits, called BS_ID, is placed in the most significant part of the flit to be transmitted, so the internal bus knows where to send it. Each routing table is statically generated and is responsible for placing this identifier in each flit. The internal bus interface follows a daisy chain topology, where each element communicates with the bus through FIFOs. The size of the FIFOs is equal to the size of each flit. For intra-FPGA communication, messages from one PN to another skip the rest of the layers, thus optimizing the latency. In the case of external communication, the bus redirects the flits to the Network Controller which sends the data outside of the FPGA.

The Network Controller implements the reliability, link and physical layers of PlasticNet. It is in charge of sending packets from one FPGA to another taking advantage of the

978-1-7281-3147-4/20 $31.00 © 2020 IEEE

maximum serial link available on each board. Each FPGA has two Network Controllers, one responsible for moving data to the FPGA located to the right and another responsible for moving data to the FPGA in the opposite direction, following a ring topology. Whenever a packet must be sent out of the FPGA, the routing tables determine the Network Controller the packet must be sent to, based on the shortest distance. A ring topology was selected since: 1) interconnection resources of the evaluation boards are sparse and 2) even though the communication is intensive, few applications describe an all-to-all interconnection architecture among the PNs that make up the network.

Within the Network Controller, the reliability layer follows a structure similar to that proposed in [12]. In this layer, a Cycle Redundancy Check (CRC) is performed on each packet, and then, a 5-bit sequencer is added to each one of them in the transmitter side, to identify the packet in an auxiliary buffer, used for re-transmitting the packet in case of an error. As soon as the CRC on the receiver side is verified, the receiver must send an acknowledge indicating which packet was received. Each sent packet in the auxiliary buffer is extracted as soon as an acknowledge is received. If the timeout assigned to receive the acknowledgment of a packet expires (either because the CRC check failed or because the sequence was broken), the transmission of new packets stops and all packets stored in the auxiliary buffer are sent again. Although this reliability layer consumes more hardware and decreases the bandwidth of physical links, applications in hardware do not allow handling errors naturally and therefore, this approach was chosen so to keep flexibility of the hardware application at the highest level.

Once the CRC has been added/checked, the data is passed through the link layer. This layer is responsible for the framing/deframing of each packet. Thus, this layer adds/removes one start and one stop byte in each packet. Then, a line code 8b10b is used to transform the data from 8 bits to 10 bits to achieve DC-balance and bounded disparity, with an efficiency of 80% (it is also possible to extend the efficiency by using a 64b66b line code). This coding provides enough state changes to allow reasonable alignment of the data stream and efficient clock recovery at the receiver side, thus achieving data transfer rates of more than 10Gbps.

Finally, the physical layer at the lowest level is responsible for serializing/deserializing the data and sending it through physical serial links. The system is flexible and can operate using single-ended signaling, low-voltage differential signalling (LVDS) (with high-speed serial links as a future option). This layer is implemented using Xilinx IPs.

VI. EVALUATION

The overhead generated in the network controller—whenever a packet is sent from one FPGA to another over the serial links—was first evaluated. Figure 2 plots the channel overhead as function of the packet size and encoding. Note that regardless of the encoding used, using packet sizes between

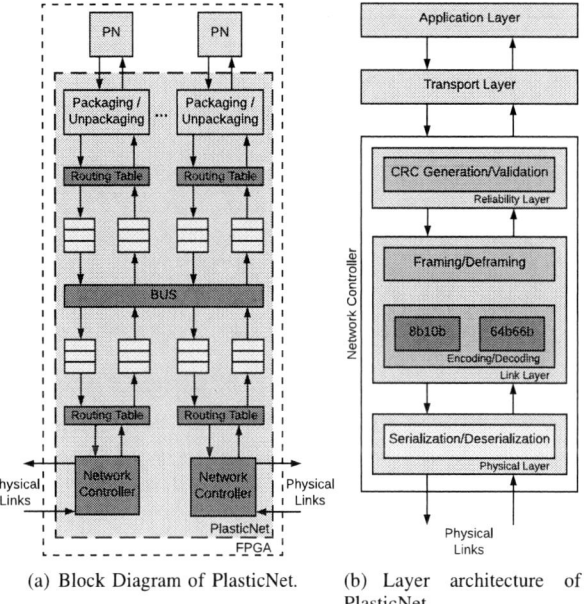

(a) Block Diagram of PlasticNet.   (b) Layer architecture of PlasticNet.

Figure 1. Description of the internal architecture of PlasticNet. Blocks or layers implemented using HLS are green-colored. Verilog blocks are represented in blue, with shades indicating different levels of abstraction. Physical layer IPs and FIFO blocks are yellow-colored and were generated using Xilinx's IP libraries.

64B and 1kB, provides an overhead similar to that one theoretically expected, without compromising the area significantly as will be seen later in Table II.

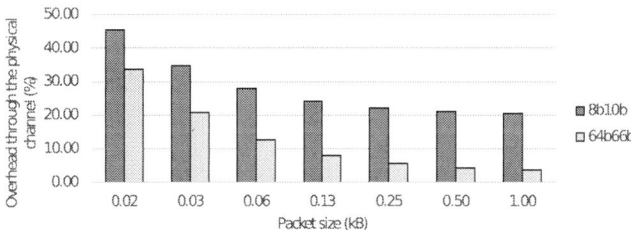

Figure 2. Channel overhead of the transmission channel, depending on the packet size and the type of encoding, assuming a CRC32 for reliability checks.

A message generator using C++ was then developed, using Xilinx Vivado HLS 2018.3, in order to test communication between PNs, and evaluating its performance on a stack of four ZedBoard Zynq®-7000 ARM/FPGA SoCs, selected primarily because of their affordability. The HLS approach allows, besides, for straightforward porting to other Zynq models The application generates different numbers of PNs that share messages among them (ranging from 1kB to 8kB in size), by sending packets from 16B to 1kB through the networking framework provided by PlasticNet. The interconnect architecture uses an 8b10b encoding and CRC32 for error detection. Additionally, each FIFO was accordingly sized to avoid congestion in the network. Table II shows a summary of resources used, based on the packet size and the number of PNs within each FPGA. Here, one can observe that the system is lightweight for packets smaller than 1024 bits. The worst-case latency for the intended application was roughly 24ms, obtained considering the exceptionally rare case where all FIFOs are full and a packet is sent to a PN located in the furthest FPGA board. Even in this very unlikely scenario, such latency may be considered as acceptable.

A half-occupancy latency scenario was then set up, which corresponds with a more likely situation where FIFOs are working at half capacity and FPGAs only connect with one another. Figure 3 plots the results of such evaluation, where one can notice that, in all the packet sizes tested, latency was less than 0.5ms, which is a good performance overall. Moreover, from the same figure, one can derive that the optimal packet size is 512B. Using this optimal packet size, PlasticNet can achieve a half-occupancy latency of only $16.9\mu s$. Just as a reference, when comparing the performance of PlasticNet against a 1Gb Ethernet network connection, results reported in [10], claim a latency of 1ms (and a $\approx 150\mu s$ ping latency) using the same Zedboard. Therefore, PlasticNet outperforms Ethernet in the same board, featuring transport control and reliability checks, but without recurring to the complexity of TCP/IP. However, when comparing PlasticNet against BlueLink, another custom bus architecture reported in [12], PlasticNet's average latency is 7.5 times slower ($16\mu s$ against $500ns$ reported in [12]). These results, nonetheless, are not discouraging, considering that in the current work only relatively low speed serial links (LVDS) per FPGA were tested (whereas high-speed multiple lanes serial interfaces are used in [12]). With the support of high-speed multiple lanes serial connections to be incorporated in future versions of PlasticNet (by integrating Xilinx's physical cores to the already available bus architecture), one would expect it to be a serious contender against BlueLink in terms of speed, but with the plus of having a much more flexible, easily integrated architecture, due to the use of a HLS methodological approach.

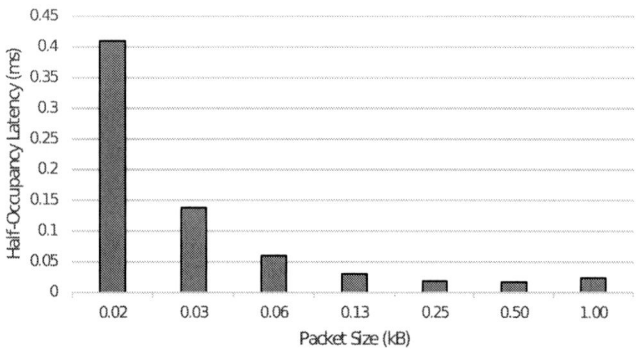

Figure 3. Half-Occupancy latency (ms) of one FPGA depending on the packet size.

VII. CONCLUSIONS

In this paper, we introduced PlasticNet, a custom network architecture for interconnected multi-FPGA system which is flexible, reconfigurable and fully compatible with applications designed using a HLS approach. PlasticNet is competitive

Table II
DEVICE UTILIZATION SUMMARY FOR PLASTICNET, TAKEN FROM XILINX'S VIVADO 2018.3 POST-SYNTHESIS REPORT USING A ZEDBOARD ZYNQ®-7000 ARM/FPGA SoC DEVELOPMENT BOARD

Number of PNs	Packet Size (bits)	128	256	512	1024	2048	4096	8192
2	Slice Registers (%)	16.50	16.98	17.94	19.87	23.72	31.42	46.81
4		28.00	30.96	32.89	33.74	41.44	56.90	85.63
2	LUTs (%)	16.67	17.27	18.48	20.88	25.70	35.32	54.57
4		27.21	28.31	30.48	34.81	43.47	60.80	90.44

in terms of overhead, latency and area against other custom networks proposed in the literature, with preliminary results showing an optimal half-occupancy latency of just $16.9\mu s$. This demonstrates that the current proposal is viable for applications with high communication requirements, with further development necessary.

Future works include: 1) incorporating support for high-speed serial links to the physical layer, 2) adding a custom line code 128b130b and, 3) exploring other interconnection topologies between boards, instead of a ring interconnect.

REFERENCES

[1] K. D. Underwood et al., "Closing the gap: CPU and FPGA trends in sustainable floating-point BLAS performance," in *12th Annual IEEE Symposium on Field-Programmable Custom Computing Machines*, April 2004, pp. 219–228.

[2] K. D. Underwood, "FPGAs vs. CPUs: trends in peak floating-point performance," in *FPGA*, 2004.

[3] K. Alfaro-Badilla et al., "Prototyping a Biologically Plausible Neuron Model on a Heterogeneous CPU-FPGA Board," in *2019 IEEE 10th Latin American Symposium on Circuits Systems (LASCAS)*, Feb 2019, pp. 5–8.

[4] K. Alfaro-Badilla et al., "Improving the simulation of biologically accurate neural networks using data flow HLS transformations on heterogeneous SoC-FPGA platforms," *High Performance Computing*, 2019, 6th Latin American Conference, CARLA 2019.

[5] Q. Tang et al., "Performance Comparison between Multi-FPGA Prototyping Platforms: Hardwired Off-the-Shelf, Cabling, and Custom," in *2014 IEEE 22nd Annual International Symposium on Field-Programmable Custom Computing Machines*, May 2014, pp. 125–132.

[6] The Dini Group - Leader in High Performance FPGA Boards. [Online]. Available: https://www.dinigroup.com/web/index.php

[7] O. Mencer et al., "Cube: A 512-FPGA cluster," in *2009 5th Southern Conference on Programmable Logic (SPL)*, April 2009, pp. 51–57.

[8] S. W. Moore et al., "Bluehive - A field-programable custom computing machine for extreme-scale real-time neural network simulation," in *2012 IEEE 20th International Symposium on Field-Programmable Custom Computing Machines*, April 2012, pp. 133–140.

[9] K. Sano et al., "Multi-FPGA Accelerator for Scalable Stencil Computation with Constant Memory Bandwidth," *IEEE Transactions on Parallel and Distributed Systems*, vol. 25, no. 3, pp. 695–705, March 2014.

[10] P. Moorthy et al., "Zedwulf: Power-Performance Tradeoffs of a 32-Node Zynq SoC Cluster," in *2015 IEEE 23rd Annual International Symposium on Field-Programmable Custom Computing Machines*, May 2015, pp. 68–75.

[11] M. Al Kadi et al., "Multi-FPGA reconfigurable system for accelerating MATLAB simulations," in *2014 24th International Conference on Field Programmable Logic and Applications (FPL)*, Sep. 2014, pp. 1–4.

[12] A. Theodore Markettos et al., "Interconnect for commodity FPGA clusters: Standardized or customized?" in *2014 24th International Conference on Field Programmable Logic and Applications (FPL)*, Sep. 2014, pp. 1–8.

Low-power activity recognition from triaxial accelerometer data

Maximiliano Chiossi
Departamento de Ingenieria
Universidad Católica del Uruguay
Montevideo, Uruguay
chiossi@gmail.com

Matias Miguez
Departamento de Ingenieria
Universidad Católica del Uruguay
Montevideo, Uruguay
mmiguez@ucu.edu.uy

Abstract—In this work, a low power triaxial accelerometer data acquisition prototype is presented, which is used to derive physical activity recognition algorithms to be used for implantable or wearable applications. There different characterization methods were implemented and can predict the different activities with good accuracy (380 cases out of 386).

Keywords—accelerometer - Low power – activity recognition

I. INTRODUCTION

Determining the physical activity that a patient is doing is crucial for several implantable medical devices (IMD) as the response required in each case can be different. In particular, rate adaptive pacemakers change heart rate to track patient's need while performing different activities. Because it is a non-invasive sensor, traditional adaptive pacemakers mostly use an accelerometer to estimate physical activity; and since energy consumption is a major concern in implantable electronics, piezoelectric sensors were traditionally utilized because of their almost null self-power consumption (note physical activity estimation is an always-on circuit block). For example, in [1][2], the integrated signal conditioning of a single-axis piezoelectric accelerometer aimed at adaptive pacemakers is presented. The sensor's signal is amplified and filtered, and then turned into a quasi-DC output proportional to the last 5-seconds average of the acceleration's amplitude in the band from 0.5 to 15 Hz The analog circuits in [1][2] consume a few hundred nA which is almost insignificant for a pacemaker, but still present limitations, e.g. to identify stairs climbing which requires a greater effort and higher heart rate, or isometric exercises. But the most relevant limitation in the case of implantable medical products in the circuits in [1][2] are ASICs and the development of an ASIC is not always possible for an IMD where production series are most of the time limited to a few thousand devices.

But recent development of commercial micro power three axes accelerometers integrating the sensor and signal conditioning also at a low cost ([3] and [4]), and probably also gyroscopes in the near future, may allow to implement physical activity recognition aimed at IMDs using off-the-shelf ICs. Also 3-axis accelerometers/gyroscopes combined with an intelligent signal processing and pattern recognition as is now possible with modern low-power microcontrollers may allow for better physical activity recognition while maintaining the low power required for IMDs. In this work, a prototype system aimed at physical activity identification in IMDs using two different commercially available low power triaxial accelerometers is presented, and different techniques for determining the activity of the patient are evaluated as a proof of concept. While probably physical activity estimation circuit embodiments using standard micropower accelerometers and microcontrollers may consume several times more power than the ASIC-based single-axis implementations in [1][2], they may allow a drastic reduction

in development time and cost, thus being an attractive option for IMDs. But even when using state-of-the-art hardware, an accurate physical activity estimation in the µW power consumption order will require intelligent and adaptive signal processing and plenty of tests with many individuals while doing different physical activities. This work is a first step towards this objective.

II. SYSTEM DESCRIPTION

A simple prototype was developed to test two different off-the-shelf accelerometers and signal processing techniques. Fig. 1 shows a block diagram of the system, while Fig. 2 shows a photograph of the working prototype. The microSD card is for simple data acquisition but would not be part of the finished product.

A. Hardware

The prototype was implemented in a 4x6 cm² dual-layer PCB. The selected microcontroller is a 16-bit

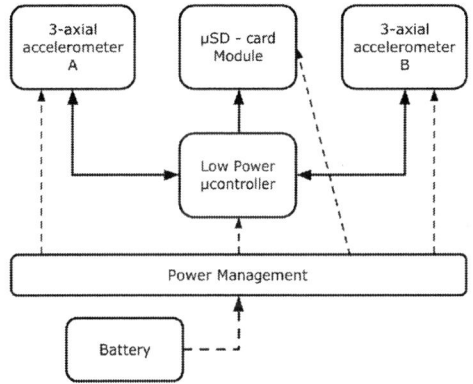

Fig. 1. Block diagram of the physical activity estimation prototype.

Fig. 2. Photograph of the implemented prototype.

978-1-7281-3147-4/20 $31.00 © 2020 IEEE

PIC24F16KL402 [5], which is a low-power microcontroller running up to 20MHz clock, programmed via ICSP. The two accelerometers under test are the ADXL362 [4] and the LIS3DH [5]. Both accelerometers provide 3 axis data at variable frequency and precision that can be selected by the microcontroller. There is a trade-off between precision and data rate with power consumption. The microSD interface is a regular off-the-shelf module with an SPI interface, as this is only to get the crude data during tests.

When designing the circuit, care was taken to align both accelerometers' measurement axes to ensure a proper comparison.

A single 3.7V LiPo battery was selected as the power source. Equivalent batteries are utilized in rechargeable IMDs, but pacemaker normally use primary batteries of Lithium iodine type with a $2.8V_{nom}$ supply voltage (beginning of life [6]), so a LDO [7] regulator was used to set a more realistic $V_{dd} = 2,7V$ supply for the microcontroller and accelerometers. A L6920DC step-up [8] was used to boost the voltage to 5V to power the microSD module [9].

B. Firmware

A simple firmware was implemented to collect data from both accelerometers; its flowchart diagram is shown in Fig. 4. The frequency selected for data sampling was firstly 12.5Hz and 10Hz for each accelerometer (as close as possible) to preserve a low power consumption. For this sampling rate, the power consumption of each accelerometer is approximately 1.8µA and 3.7µA respectively.

The microcontroller handles the configuration of the accelerometers and uses rising-edge external interrupts to read the 16-bit data output of the accelerometers, storing the formatted data on a text file on the microSD card with a FAT32 filesystem for easy access.

A sampling run is started with a push button, and with each subsequent interrupt a LED provides visual feedback to the user. Pressing the push button again stops the sampling process and saves the data onto the microSD card for later analysis.

C. Activity estimation setup

Because the objective is to measure the overall patient's physical activity, a location close to the center of mass is usually desirable.

For the tests in this work the module was installed on the right side of the hip [10] as shown in Fig. 3, in a healthy individual. With this configuration, the accelerometers' X axes point upwards, the Y axes point backwards, and the Z axes point to the right. The prototype's reduced size and weight allows for a comfortable wear while taking measurements.

III. EXPERIMENTAL MEASUREMENTS

Experimental data was obtained from one subject, recording five common activities that are interesting to discriminate: sitting down, lying down, walking, and climbing up and downstairs. More than 50 samples, each of a 6-second sampling window, were obtained for each activity. The exact conditions of the measurements will be described in this section.

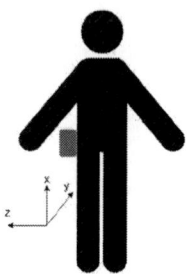

Fig. 3. Prototype placement for the tests.

A. Sitting down

The subject sat down on an office chair while carrying out office work on a computer. The amount of activity is limited to a minimal, with sporadic movements of the body (arms, hips), which keeps the acceleration measurements almost restricted to measuring the Earth's gravity, serving as a baseline for the rest of the experiments.

B. Lying down

Similar to the sitting down experiment, the subject lay down facing upwards on a bed and had a minimal activity. The subject remained lying down on a bed while using a cellphone, which requires minor movement and effort.

C. Walking

While walking can sometimes require considerable effort, for this experiment the subject walked at a normal pace on flat ground, with no obstacles on the way.

D. Climbing up and downstairs

For this experiment the subject took turns at climbing up and down a single 12-step stair. The measurements were taken continuously climbing up and down, starting on the way up, and each subsequent run was separated with a 2-3 second wait

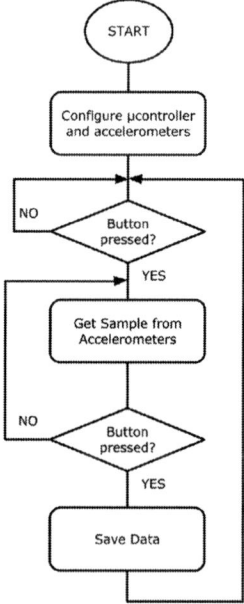

Fig. 4. Simplified diagram of the implemented firmware for data collection.

978-1-7281-3147-4/20 $31.00 © 2020 IEEE

that made it easy to then identify each climb during the signal processing stage.

IV. ACTIVITY DETERMINATION

The measured physical activity was processed using 6 second intervals, which correspond to 60 samples for the LIS3DH and 75 samples for the ADXL362. Each of these packets was analyzed to extract the parameters to use for the classification algorithm. The main processing was performed off-line using Scilab [11]. When comparing the data from the different accelerometers, no significant difference was obtained, concluding that any of the chosen accelerometers can be used in this application. Even though the LIS3DH is cheaper, its larger power consumption makes the ADXL362 the better candidate for this application.

To determine the activity, we will develop a series of machine learning algorithms that can infer the activity from a group of parameters that can be extracted from the acceleration data.

A. Parameter extraction

Several different parameters were extracted for each data packet trying to minimize computational effort, as the final application should be performed by a low power microcontroller. The following characteristics were extracted following a procedure similar to [10]:

$$SMA_X = \frac{\sum_{i=1}^{n}|x_i|}{n} \; ; SMA_Y = \frac{\sum_{i=1}^{n}|y_i|}{n} \; ; SMA_Z = \frac{\sum_{i=1}^{n}|z_i|}{n} \tag{1}$$

$$E_x = \frac{\sum_{i=1}^{n}x_i^2}{n} \; ; \; E_y = \frac{\sum_{i=1}^{n}y_i^2}{n} \; ; \; E_z = \frac{\sum_{i=1}^{n}z_i^2}{n} \tag{2}$$

$$\bar{x} = \frac{\sum_{i=1}^{n}x_i}{n} \; ; \; \bar{y} = \frac{\sum_{i=1}^{n}y_i}{n} \; ; \; \bar{z} = \frac{\sum_{i=1}^{n}z_i}{n} \tag{3}$$

$$\sigma^2{}_x = \frac{\sum_{i=1}^{n}(x_i-\bar{x})^2}{n-1} \; ; \; \sigma^2{}_y = \frac{\sum_{i=1}^{n}(y_i-\bar{y})^2}{n-1} \; ; \; \sigma^2{}_z = \frac{\sum_{i=1}^{n}(z_i-\bar{z})^2}{n-1} \tag{4}$$

In equations 1-4, the symbols x, y, and z represent the samples of each axis of the accelerometer. SMA in (1) is the signal magnitude area, E in (2) is the energy of the signal, and finally (3) is the average and (4) is the variance of each signal.

The 12 parameters were calculated, four parameters per axis. From these characteristics, we would like to select which are the most important to implement in the microcontroller.

B. Visualization and Classification

To visualize and apply different classification methods, the Orange Data Mining Python software [12] was used.

Fig. 5 shows a dispersion diagram using just two variables, (SMA_Y and $\sigma^2{}_X$) which already shows a good discrimination between the different activities and which information is stored in each parameter. Fig. 6 shows a linear projection using 4 parameters, which further separates the different activities into different clusters.

Three different simple classification methods were tested: Logistic Regression [13], Support Vector Machine (SVM) [13], and nearest neighbors (kNN) [13]. Using the experimental data, the models were trained using 10-fold cross validation, and their performances analyzed via their confusion matrices [14]. In tables I-III, the confusion matrices are presented for each of the algorithms.

TABLE I.   LOGISTIC REGRESSION

Actual	Predicted				
	Down	Lying down	Sitting	Up	Walking
Down	54	0	0	3	0
Lying down	0	64	0	0	0
Sitting	0	0	68	0	0
Up	3	0	0	53	0
Walking	0	0	0	0	141

TABLE II.   SVM

Actual	Predicted				
	Down	Lying down	Sitting	Up	Walking
Down	53	0	0	4	0
Lying down	0	64	0	0	0
Sitting	0	0	68	0	0
Up	2	0	0	54	0
Walking	0	0	0	0	141

TABLE III.   kNN

Actual	Predicted				
	Down	Lying down	Sitting	Up	Walking
Down	52	0	0	5	0
Lying down	0	64	0	0	0
Sitting	0	0	68	0	0
Up	7	0	0	49	0
Walking	0	0	0	0	141

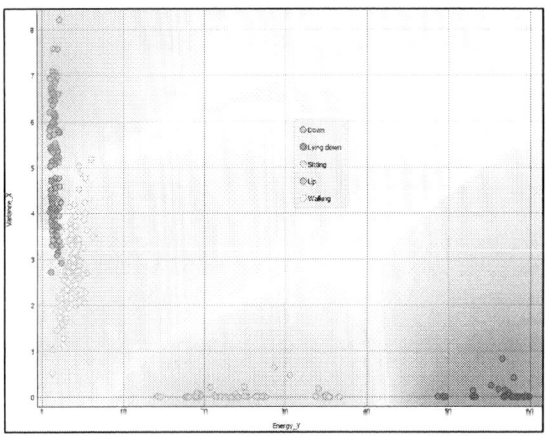

Fig. 5.   Dispersion diagram using two variables (E_Y horizontal, $\sigma^2{}_X$ vertical). Blue = down, red = lying down, green = sitting, brown = up, yellow = walking.

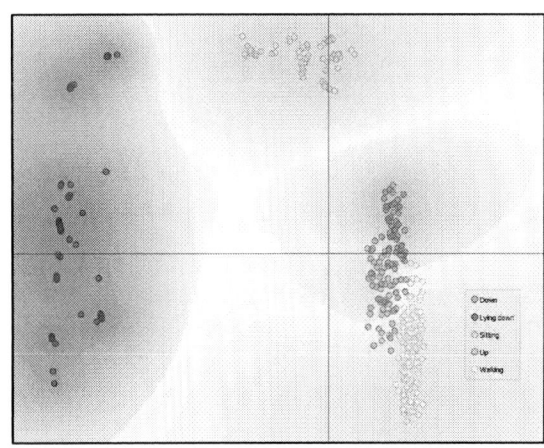

Fig. 6.   Linear projection using four parameters (SMA_X, SMA_Z, E_Z, $\sigma^2{}_X$). Blue = down, red = lying down, green = sitting, brown = up, yellow = walking.

C. Results

The results from the three algorithms are virtually similar. All of them detect the difference between lying down, sitting, walking and going up and down the stairs. There are a few identification errors (6/386 for Logistic regression and SVM while 12/386 for kNN) but they always occur between going up and downstairs. The kNN methods is slightly worse at predicting whether the person is climbing up or down a stair but all methods are excellent in determining if the person is walking, sitting, lying down or climbing a stair.

V. CONCLUSIONS

A first prototype for activity recognition using a commercial triaxial accelerometer was developed. Two different accelerometers were tested, yielding almost identical results. The data collected was analyzed and using some simple parameters that can be easily extracted, three different classification method were tested, which show excellent results.

The next step in this project is implementing the classification method in the microcontroller. To this end, the calculations for each of the algorithms from tables I through III have to be coded into the microcontroller as done in the Orange Data Mining software. We will pay special attention to Logistic Regression, as it requires the least amount of computational effort, therefore requiring less power.

Further consumption reduction can be achieved in a number of ways, such as adjusting the clock speed and microcontroller sleep cycles during samples, as well as putting the accelerometer asleep for longer periods under certain conditions, for example, if the subject has been lying down for a reasonably long period (which would equate to assuming the subject is asleep and will remain so for an extended period, and therefore requiring no frequent measurements).

The ADXL362 accelerometer has the capability of generating interrupts when the acceleration detected is above a certain threshold, which can be used to detect when each activity exceeds a certain level of intensity (the threshold will likely depend on each activity and each individual), allowing the microcontroller to remain asleep for longer and further reduce power consumption until the intensity of the activities carried out is of interest.

REFERENCES

[1] A. Arnaud and C. Galup-Montoro, "Fully integrated signal conditioning of an accelerometer for implantable pacemakers," J. Analog Integr. Circuits Signal Process., vol. 49, pp. 313–321, Dec. 2006.

[2] J. Gak, M. R. Miguez and A. Arnaud, "Nanopower OTAs With Improved Linearity and Low Input Offset Using Bulk Degeneration," in *IEEE Transactions on Circuits and Systems I: Regular Papers*, vol. 61, no. 3, pp. 689-698, March 2014.

[3] ADXL362: https://www.analog.com/en/products/adxl362.html

[4] LIS3DH: https://www.st.com/en/mems-and-sensors/lis3dh.html

[5] PIC24F16KL402: https://www.microchip.com/wwwproducts/en/PIC24F16KL402

[6] Mallela, Venkateswara Sarma et al. "Trends in cardiac pacemaker batteries." *Indian pacing and electrophysiology journal* vol. 4,4 201-12. 1 Oct. 2004

[7] TCR2EF: https://toshiba.semicon-storage.com/us/product/linear/power-supply/detail.TCR2EF27.html

[8] L6920DC: https://www.st.com/en/power-management/l6920dc.html

[9] MicroSD module: https://www.amazon.com/Module-Storage-Adapter-Interface-Arduino/dp/B07PFDFPPC

[10] I. Farkas, E. Doran, "Activity Recognition from Acceleration Data Collected with a Tri-axial Accelerometer", *Acta Tech. Napocensis-Electron. Telecommun.*, vol. 52, no. 2, pp. 38-43, 2011.

[11] Scilab: https://www.scilab.org/

[12] Orange Data Mining: https://orange.biolab.si/

[13] Richard O. Duda, Peter E. Hart, and David G. Stork. 2000. Pattern Classification (2nd Edition). Wiley-Interscience, New York, NY, USA.

[14] Stephen V. Stehman, "Selecting and interpreting measures of thematic classification accuracy" Remote Sensing of Environment, Volume 62, Issue 1, 1997, Pages 77-89

Ultra-Low Power Relaxation Oscillators survey: Design Trends and Challenges

William Teles Medeiros[1], Hamilton Klimach[1,2], Sergio Bampi[1]

[1]Microelectronics Graduate Program - UFRGS - Porto Alegre, RS, Brazil

[2]Electrical Engineering Dept. - UFRGS - Porto Alegre, RS, Brazil

william.meds@gmail.com, hamilton.klimach@ufrgs.br, bampi@inf.ufrgs.br

Abstract—The Relaxation oscillators (ROSCs) offer a better trade-off between power consumption, physical area and accuracy. This paper presents a study of recently published ultra-low-power relaxation oscillators with low-temperature coefficient, identifying its characteristics, performance parameters, and main challenges. A brief review show as state-of-the-art circuits handle with PVT (Process, voltage, and temperature) variations. By the end of this analysis, some trends in relaxation oscillators design are discussed.

Index Terms − **Ultra-low power consumption, relaxation oscillator, PVT variations.**

I. Introduction

Ultra-Low Power devices must be able to receive information, process data, and transmit become essential to the smart objects [1]. Many Integrated Circuits (IC) applications such, radios, sensors, power management, and interface required an accurate clock reference to work properly. Crystal oscillators offer excellent frequency stability in respect of process, supply voltage, and temperature (PVT) variations [2]. They are used in KHz-MHz range, providing a frequency accurate at the cost of power consumption, larger physical size (Off-Chip) and highly expensive. One of the numerous sub-circuits in the Internet of Things (IoT) devices is the Real-Time Clock (RTC) that must stay awake all the time, even when other sub-circuits are in sleep mode. Therefore, it must consume the minimum as possible [3]. An oscillator can be employed to generate a stable frequency required by RTC to work properly. In an oscillator design, there are trade-offs between size, power efficiency and accuracy. Usually, LC oscillators consume a large area, high power consumption, which is not a good option for IoT applications. There are a few types of oscillators with low power consumption enough to be employed in IoT applications [4]. RC oscillators, including ring oscillators, wien-bridge oscillators, and relaxation oscillators, have been widely studied and developed for Systems-on-Chip (SoC) due to their high compatibility with the standard CMOS process and their small physical area [5].

In general situation, wien-bridge oscillators and ring oscillators can be designed with a small area and low power consumption. However, they suffer more with frequency variation than relaxation oscillators. Relaxation oscillators offer a good trade-off between physical area, power consumption, start-up time and temperature sensitivity [6].

It makes relaxation oscillator the most commonly used for ultra-low power on-chip applications. However, these benefits come at the expense of higher clock jitter, which is not a key concern for many low power applications [7].

In this work, a review of state-of-the-art relaxation oscillators is presented, and we discuss initial design trends and challenges. The paper is organized as follows: Section II summarizes the main characteristics and performances of relaxation oscillators. In section III presents a review, strategies, and results are discussed. Section IV presents the concluding remarks.

II. Relaxation Oscillator Characteristics and Performance Parameters

A conventional relaxation oscillator presented by [8], [9], [10], [2], [11], [12], [13] and [7] contains two comparators, a RS Latch, a bias reference and timing circuit. The operation principle of the relaxation oscillator occurs through charging and discharging of capacitors. The continuous comparator compares the capacitor voltage with the reference voltage, and when the capacitor voltage achieves the reference voltage, the comparator produces a pulse that is maintained by the RS Latch until the next comparison. The Latch output signal controls the switches responsible by charge or discharge of the capacitors.

This topology shown in Fig. 1 (a) works in two phases: The charging phase of C1, shown in Fig. 1 (b) and discharging phase, shown in Fig. 1 (c). The capacitor C1 is charging through the switch S2 by I_c, while capacitor C2 discharge through S3 to ground. At the moment when the capacitor voltage VC1 achieves V_{REF} voltage, the output of the comparator COMP1 changes to set Q high and leads to the discharge phase of C1. In the discharge phase, the capacitor C2 is charged through the switch S4 by I_c, while capacitor C1 is discharging through S1 to ground. When the capacitor voltage VC2 achieves the V_{REF} voltage, the output of the comparator COMP2 changes to set QB to high, and Q is set to low restarting the charge phase [9].

Oscillation frequency, power consumption, and temperature coefficient (TC) are usually the most relevant performances of a relaxation oscillator. However, other important parameters, such as line sensitivity, long term stability, and start-up time, have importance as well. The frequency deviation due to the temperature variation is the most critical specification for a

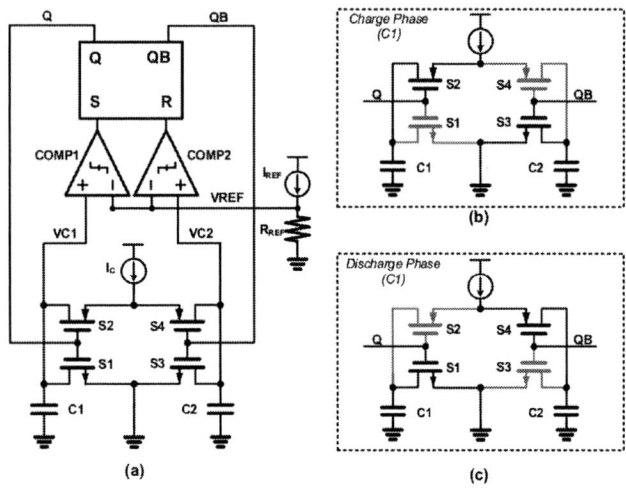

Fig. 1. Conventional Relaxation Oscillator [9].

ROSC when used in ICs that demand precision clock. This performance is called the temperature coefficient and takes into account the temperature range as well. Line sensitivity is a metric used to measure the influence of the supply voltage variation in the oscillation frequency. The temperature coefficient and line sensitivity can be directly compared with each other because both measure instant frequency change due to short-term changes. Another metric of frequency deviation is long-term stability, commonly called Allan deviation floor. It is often used to evaluate the noise performance on oscillators. As averaging time increases, frequency fluctuations decrease as white noise is averaged out until flicker noise dominates [14]. Start-up time is the amount of time required until the oscillator reaches the steady-state, stabilizing at the desired operating frequency.

Ideally, RC values determine the nominal frequency of a ROSC. The main problems that arise in the ROSC are offset voltage, delay time, leakage and tunneling current, and current mismatch. Comparator offset and delay time across PVT variations have been identified as a major bottleneck for low power relaxation oscillators [15]. The problem is not the average values of these effects but its behavior with PVT variation that affects accuracy.

In the next section, we will discuss which techniques are being implemented to minimize the effects shown above.

III. REVIEW AND DISCUSSION

This paper aims to review and identify the main design issues related to relaxation oscillators development. Table I offers a summary of state-of-the-art ROSCs, highlighting their main performance figures.

The Offset voltage appears due mainly to the mismatch between par-differential transistors in the comparator. The transistors can be sized to minimize the offset voltage. Usually, offset voltage increases linearly with temperature variation and vary slightly with the voltage supply. The author [14]

proposes a relaxation oscillator working in two phases and presents an offset cancellation that uses a timing circuit, comparator, and Schmitt trigger. Offset cancellation happens due to the reference voltage stored in a capacitor, switches in the input of the comparator. It is connected, positive input in half period and negative input in the other period. The offset voltage of the comparator ends up being transferred to the duty cycle variation of the output signal, which does not affect the accuracy of the oscillator. In the architecture proposes by [22], a conventional continuous comparator is replaced by a pseudo-inverter chain to avoid comparator offset effect. The architecture consists of a timing circuit, a source follower, and an inverter chain.

Delay time is usually not a problem at low frequencies when power consumption is not a problem. Therefore, when we are trying to optimize power consumption for ULP systems, delay time becomes a great issue. The comparator delay mostly defines the delay of the whole circuit. The delay time is also sensitive to temperature and voltage supply variation. It exhibits nonlinear behavior in temperature range, and it is directly associated with the bandwidth of the comparator. The ROSC proposed by [9] presents a conventional relaxation oscillator with a half-period pre-charge compensation scheme. There are four stages in a period of the proposed oscillator: normal charge, discharge, pre-charge and hold. By controlling the pre-charge time, the capacitor is charged to the same voltage value as the delay time adds to the period. Thus, the delay time influence on the oscillation frequency is eliminated. The topology is similar to Fig. 1 and includes two 2-to-1 multiplexers, extra switches, and a control logic generation. A current reference was employed by [14] to generates a proportional to absolute temperature (PTAT) that was used as a bias to the comparator and timing circuit to compensate the delay time. A PTAT current biasing leads a constant bandwidth (constant-gm) across temperature, and consequently, constant delay.

A different approach for RC oscillators is proposed in [17] that introduces a resistive frequency locked loop topology for accurate clock generation. In this topology, a switched capacitor is controlled by a voltage controlled oscillator (VCO). The basic principle is to generate a stable frequency by matching the equivalent resistance of the switched capacitor circuit to a temperature compensated on-chip reference resistor using an ultra-low-power amplifier. The rate of the control signal for the switched capacitor is the output frequency of this oscillator. This approach eliminates the comparator that limits the temperature stability due to your temperature-dependent delay. The amplifier which replaces the comparator can be low-bandwidth and ultra-low power. A new architecture developed by [20] proposes a relaxation oscillator employing a two-stage comparator along with a pseudo-differential amplifier (PD-AMP) in feedback to improve PTAT current temperature linearity. The key takeaway from this analysis is that the comparator delay consists of two terms: one relating to the finite bandwidth (τ_{BW}) and another dependent on the following buffer stage switching threshold (τ_{SW}). The τ_{BW}

978-1-7281-3147-4/20 $31.00 © 2020 IEEE

TABLE I
RECENT RC OSCILLATORS IN RELEVANT PUBLICATIONS

Work	Process	Supply [V]	Frequency [KHz]	Temp. Coef [ppm/$^{\circ}C$]	Freq. Variation [%]	Temp. Range [$^{\circ}C$]	Line Sens. [%/V]	Long Term Stab. [ppm]	Power [nW]	Start-up Time	Area [mm^2]	Samples
[3]	180nm	1.0	32.55	120	±0.84	-40 to 100	1.1	-	472	4 cycles	0.105	20
[9]	180nm	0.6	32.7	43.1	±0.39	-55 to 125	2.4	-	51	1 cycle	0.048	-
[14]	65nm	1.0	18.5	46.3	± 0.3	-40 to 90	<5	20	130	4 cycles	0.032	4
[15]	180nm	1.5	12	31	±0.2	-40 to 90	0.85	-	120	-	0.105	-
[16]	65nm	-	33	38.2	±0.21	-20 to 90	0.09	<4	190	-	0.015	5
[17]	180nm	1.0	70.4	34.3	±0.2	-40 to 80	0.75	<7	110	<176 cycles	0.26	5
[18]	180nm	0.7-1.8	32	30	±0.15	-20 to 80	0.05	500	150	-	0.1	5
[19]	180nm	1.2	3	13.8	±0.07	-25 to 85	0.48	<63	4.7	-	0.5	-
[20]	180nm	0.4	1.22	94	±0.42	-20 to 70	17.2	58	1.14	-	0.2	5
[21]	180nm	1.2	28	95.5	±0.48	-20 to 80	-	-	40	-	0.16	5

delay exhibits a CTAT dependence with slightly larger than the first-order. It can be counteracted by designing a linear PTAT τ_{RC} (from RC timing) by trimming R to be PTAT. To cancel τ_{SW}, a common source (CS) stage is added to the existing comparator. As a current-starved inverter, the delay of the CS stage is inversely proportional to the bias current.

The Comparator non-idealities, such as delay time and offset, can be compensated together. The ROSC designed by [3] consists of two pairs of clock and voltage reference generators, two comparators, a bias circuit, multiplexers, and a control logic circuit. Multiplexers are used to switch the connection in the input of the comparators, similar to the [9] architecture. After four operation phases, offset and delay are suppressed. The circuit proposed by [21] contains a curvature current source, a PTAT current reference, current-mode comparator, and clock buffers. A PTAT low-temperature coefficient current reference and a curvature current source are realized by transistors with different gate-oxide thicknesses. The curvature current source compensates for the curvature effect to further reduce the TC of the oscillator. Another different topology is implemented by [16]. It consists of an RC network and stages of three and five inverters, similar to the ring oscillator topology. A simple regulator that employed an NMOS voltage follower and a replica inverter biased by a 25 nA PTAT reference current produces a local regulated supply that tracks threshold voltage for the inverters.

The leakage current has a high impact on the switches (timing circuit), mainly in advanced CMOS technologies. Leakage current may provide a pre-charge or an initial discharge of capacitors in the wrong phase of operation, and it presents nonlinear behavior. Then, [14] replaces standard transistors by "high voltage" ones (1.8 V), making leakage current negligible. The tunneling current is very impactful when we are working on advanced technologies (below 65 nm) for ULP applications. Then, for advanced CMOS nodes, a trade-off arises between tunneling current and leakage current. Consequently, if it was allowed tunneling current, power consumption increases as well. Both leakage and tunneling current can be eliminated using "high voltage" transistors.

The voltage reference circuit from [9], [17], [14], [15] and [21] combines a polysilicon resistor (with a positive temperature coefficient) and a diffusion resistor (with a nega-

tive temperature coefficient) to cancel their temperature dependencies. A Zero V_T MOSFET was used as a resistor in the voltage-controlled current source [18]. The nonlinear MOS capacitor operates in the strong inversion region. The mobility variation with temperature of the Zero V_T MOSFET compensates the thermal voltage to achieve a frequency that is almost insensitive to temperature variation. Metal-insulator-Metal (MIM) and Metal-oxide-metal (MOM) capacitors was employed by [14], [17], [15] and [21] because they have a negligible temperature coefficient. The relaxation oscillator proposes by [18] replaces the conventional capacitor for a PMOS transistor used as a capacitor.

Secondary issues, which still affect the oscillator stability, can be produced by mismatch in the reference resistor, timing circuit and current sources. Another problem is that to generate a pre-defined reference voltage of few mV, with a reference current of few nA, resistors of MΩ are required. It is increasing the physical area significantly for ROSC with ultra-low power consumption. In 2016 [19] propose a self-biased wake-up timer using a switched-resistor scheme. However, to handle with problem of large resistor size, a duty-cycled resistor scheme was employed to increase resistance without increasing area. The current reuse and self-biasing techniques are also used to save power and to ensure stable operation over PVT variations. The VCO controls the switching of the switched resistor and the switched capacitor. When the current of the switched capacitor becomes equal to the switched resistor current, the frequency is locked. To reduce mismatch between two current sources, [19] proposes to alternate their connections to each input node of the amplifier, while the VCO output control this alternation. As a result of this chopping scheme, the frequency errors caused by current mismatch is removed.

The process compensation is related through manual calibration by [20], where a resistor is implemented as a resistor digital-to-analog converter. With a series combination of 5 bits PTAT and CTAT resistors to minimize the process variation. However, [17] trimmed the ratio between two resistors through 2-point on-chip after fabrication to compensate the process variation. To control and compensate for oscillation frequency characteristics, [3] employed a digitally controlled trimming.

As shown in Fig. 2, Figure of Merit (FoM) brings the

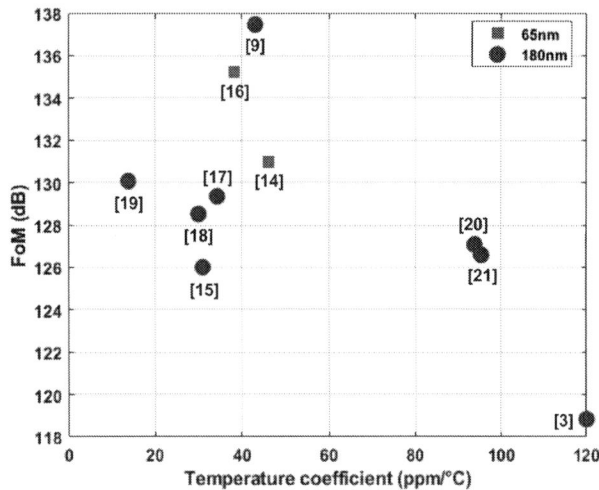

Fig. 2. Figure of Merit vs Temperature Coefficient.

relation between those figures of performance for the evaluated state-of-the-art works. We notice a behavior trend for all works, and it seems rough to achieve results better than 30 $ppm/°C$. [19] delivers the best results for TC, but it is the most complex circuit with the largest physical area, and not to mention the number of samples measured. Most circuits have a TC between 30 $ppm/°C$ and 50 $ppm/°C$, with power consumption from 51 nW to 190 nW. From the results presented, it does not appear to have significant gain or loss of performance on circuits developed in 65 nm. However, they have the smallest physical areas. Presenting a TC between 90 $ppm/°C$ and 120 $ppm/°C$, [20] has the lower power consumption, operation frequency and temperature range. [3] has the worse FoM results, because it presented the worst TC and power consumption values, but measured 20 samples.

IV. CONCLUSION

This paper presented a survey of recently published relaxation oscillators in significant publications. After a brief fundamental operation review and issues caused by PVT variations, we exploited some techniques design choices and their influence on circuits performances.

To do so, we evaluate recent works from relevant publications. The table I offers a summary of state-of-the-art ROSCs. We notice a behavior trend for all works that there is a trade-off between power, temperature coefficient, and area. However, if performance improvement is achieved in these three metrics, we always have a loss of performance on line sensitivity, long-term stability, or start-up time.

V. ACKNOWLEDGMENT

This work was supported by CNPq, CAPES and UFRGS.

REFERENCES

[1] G. Fortino and A. Liotta, *"Internet of Things Based on Smart Objects: Technology, Middleware and Applications"*. Springer, 2014.

[2] J. Wang and W. L. Goh, "A 13.5-mhz relaxation oscillator with ±0.5% temperature stability for rfid application," in *2016 IEEE International Symposium on Circuits and Systems (ISCAS)*, May 2016, pp. 2431–2434.

[3] K. Tsubaki, T. Hirose, N. Kuroki, and M. Numa, "A 32.55-khz, 472-nw, 120ppm/°c, fully on-chip, variation tolerant cmos relaxation oscillator for a real-time clock application," in *2013 Proceedings of the ESSCIRC (ESSCIRC)*, Sept 2013, pp. 315–318.

[4] S. Jeong, I. Lee, D. Blaauw, and D. Sylvester, "A 5.8 nw cmos wake-up timer for ultra-low-power wireless applications," *IEEE Journal of Solid-State Circuits*, vol. 50, no. 8, pp. 1754–1763, Aug 2015.

[5] H. Abbasizadeh, B. S. Rikan, and K. Lee, "A fully on-chip 25mhz pvt-compensation cmos relaxation oscillator," in *IFIP/IEEE International Conference on Very Large Scale Integration (VLSI-SoC)*, Oct 2015, pp. 241–245.

[6] S. Dai and J. K. Rosenstein, "A 14.4nw 122khz dual-phase current-mode relaxation oscillator for near-zero-power sensors," in *2015 IEEE Custom Integrated Circuits Conference (CICC)*, Sept 2015, pp. 1–4.

[7] B. Cimbili, D. Wang, R. C. Zhang, X. L. Tan, and P. K. Chan, "A pvt-tolerant relaxation oscillator in 65nm cmos," in *2016 IEEE Region 10 Conference (TENCON)*, Nov 2016, pp. 2315–2318.

[8] K. Choe, O. D. Bernal, D. Nuttman, and M. Je, "A precision relaxation oscillator with a self-clocked offset-cancellation scheme for implantable biomedical socs," in *2009 IEEE International Solid-State Circuits Conference - Digest of Technical Papers*, Feb 2009, pp. 402–403,403a.

[9] Y. Zheng, L. Zhou, F. Tian, M. He, and H. Liao, "A 51-nw 32.7-khz cmos relaxation oscillator with half-period pre-charge compensation scheme for ultra-low power systems," in *2016 IEEE International Symposium on Circuits and Systems (ISCAS)*, May 2016, pp. 830–833.

[10] T. Tokairin, K. Nose, K. Takeda, K. Noguchi, T. Maeda, K. Kawai, and M. Mizuno, "A 280nw, 100khz, 1-cycle start-up time, on-chip cmos relaxation oscillator employing a feedforward period control scheme," in *2012 Symposium on VLSI Circuits (VLSIC)*, June 2012, pp. 16–17.

[11] K. Tsubaki, T. Hirose, Y. Osaki, S. Shiga, N. Kuroki, and M. Numa, "A 6.66-khz, 940-nw, 56ppm/°c, fully on-chip pvt variation tolerant cmos relaxation oscillator," in *2012 19th IEEE International Conference on Electronics, Circuits, and Systems (ICECS 2012)*, Dec 2012, pp. 97–100.

[12] J. Mikulić, G. Schatzberger, and A. Barić, "A 1-mhz on-chip relaxation oscillator with comparator delay cancelation," in *ESSCIRC 2017 - 43rd IEEE European Solid State Circuits Conference*, Sept 2017, pp. 95–98.

[13] J. Wang, W. L. Goh, X. Liu, and J. Zhou, "A 12.77-mhz 31 ppm/°c on-chip rc relaxation oscillator with digital compensation technique," *IEEE Transactions on Circuits and Systems I: Regular Papers*, vol. 63, no. 11, pp. 1816–1824, Nov 2016.

[14] A. Paidimarri, D. Griffith, A. Wang, G. Burra, and A. P. Chandrakasan, "An rc oscillator with comparator offset cancellation," *IEEE Journal of Solid-State Circuits*, vol. 51, no. 8, pp. 1866–1877, Aug 2016.

[15] Y. Ma, D. Wang, S. Zhang, and X. Fan, "Integrated relaxation oscillator with no comparator for energy harvesting," *Electronics Letters*, vol. 53, no. 12, pp. 800–802, 2017.

[16] D. Griffith, P. T. Røine, J. Murdock, and R. Smith, "17.8 a 190nw 33khz rc oscillator with ±0.21in *2014 IEEE International Solid-State Circuits Conference Digest of Technical Papers (ISSCC)*, Feb 2014, pp. 300–301.

[17] M. Choi, T. Jang, S. Bang, Y. Shi, D. Blaauw, and D. Sylvester, "A 110 nw resistive frequency locked on-chip oscillator with 34.3 ppm/°c temperature stability for system-on-chip designs," *IEEE Journal of Solid-State Circuits*, vol. 51, no. 9, pp. 2106–2118, Sept 2016.

[18] A. W. Zomagboguelou, C. G. Montoro, and M. C. Schneider, "A 150nw 32 khz mobility-compensated relaxation oscillator with ±30ppm/c temperature stability," in *2016 IEEE 7th Latin American Symposium on Circuits Systems (LASCAS)*, Feb 2016, pp. 387–390.

[19] T. Jang, M. Choi, S. Jeong, S. Bang, D. Sylvester, and D. Blaauw, "5.8 a 4.7nw 13.8ppm/°c self-biased wakeup timer using a switched-resistor scheme," in *2016 IEEE International Solid-State Circuits Conference (ISSCC)*, Jan 2016, pp. 102–103.

[20] H. Jiang, P. P. Wang, P. P. Mercier, and D. A. Hall, "A 0.4-v 0.93-nw/khz relaxation oscillator exploiting comparator temperature-dependent delay to achieve 94-ppm/°c stability," *IEEE Journal of Solid-State Circuits*, vol. 53, no. 10, pp. 3004–3011, Oct 2018.

[21] Y. Chiang and S. Liu, "Nanopower cmos relaxation oscillators with sub-100ppm/°Ctemperature coefficient," *IEEE Transactions on Circuits and Systems II: Express Briefs*, vol. 61, no. 9, pp. 661–665, Sept 2014.

[22] Y. Ma, K. Cui, S. Fang, and X. Fan, "On-chip dual-phase charge-transfer relaxation oscillator with comparator offset cancellation," *Electronics Letters*, vol. 54, no. 1, pp. 23–25, 2018.

Pre-Synthesis Evaluation of Digital Bus Micro-Architectures

R. Garcia-Ramirez*, A. Chacon-Rodriguez*, C. Strydis[†], and R. Rimolo-Donadio*

*Escuela de Ingeniería Electrónica, Instituto Tecnológico de Costa Rica
[†]Dept. of Neuroscience, Erasmus Medical Center, Rotterdam, The Netherlands
Email: {rgarcia, alchacon, rrimolo}@tec.ac.cr, c.strydis@erasmusmc.nl

Abstract—Buses are central building blocks in the architecture of digital systems. There are numerous standards for bus architectures and evaluation metrics in terms of data transfer rate, quality of service, and latency; however, it is not common to find metrics related to the physical features of bus implementations, such as power consumption and area in terms of their micro-architecture. This paper evaluate bus micro-architectures at pre-synthesis level, allowing for the comparison of alternative circuits implementing the same standard and thus providing estimations on the power consumption and area requirements. A metric is proposed to evaluate the bus implementation and its utilization is shown with generic serial and parallel buses, based on simulations with a 0.18μm CMOS standard cell library.

Index Terms—Bus, Interconnects, Micro-Architecture, System-on-Chip, Very Large Scale Integration.

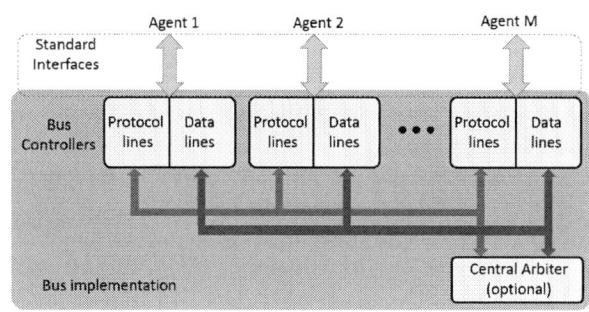

Fig. 1. Block diagram of a generic bus implementation.

I. INTRODUCTION

Buses are the preferred interconnect architecture for the implementation of digital systems; there are many variations in terms of protocols and interfaces, for instance AMBA, STbus, Avalon, Core Connect, and Wishbone, to name some typically found in modern System-on-Chip (SoC) solutions [1].

The performance of interconnect networks is often evaluated by metrics such as latency, bandwidth, and throughput [2]. Latency is defined as the average time required by the packages of information to reach their final destination. Bandwidth is the amount data per unit time that travels through the bus, and throughput is the rate of data that is transmitted between entities capable to generate and consume data (here on referred as agents), which is lower than the bandwidth. Other metrics are intended to measure quality of service and fault tolerance.

All the previous metrics are affected by the protocol stack used, which defines the conformation of the flit[1], as well as by the general architecture of the bus [1]. Once these are fixed, the bus micro-architecture becomes the main factor determining its final power consumption and area. As such, in order to evaluate in detail the area and power of any particular micro-architectural implementation, it is necessary to carry out its design at least to the gate level.

Thus, it would be useful to have metrics allowing for the micro-architectural evaluation in an early stage, without requiring a full synthesis of a particular implementation. Analytical

[1]Defined as the minimum package of information which can travel in the physical layer of the bus implementation

models intended to the analysis of delay [3], throughput [4] in Network on Chip (NoC), and methodologies for the evaluation of power at the CPU level [5] can be found in literature; but there are no metrics that can be used for the evaluation of the complexity of a micro-architecture for a given bus architecture and stack of communication protocols.

This paper proposes a general metric for the evaluation of bus implementations. The proposed approach is applied for the evaluation of serial and parallel buses, which are compared in terms of area and power requirements. Metrics are validated against behavioral simulations using libraries of a 0.18μm CMOS process.

II. METRIC PROPOSAL

A basic representation of the implementation of a bus interconnect is presented in Fig. 1. Every bus consists of bus controllers connected to a shared medium; the standard interface defined by the architecture is reflected in the connection between the bus controller and the agents. The shared lines connected to all the bus controllers can be classified as data lines, used for the transmission of payload, and control lines, intended for synchronization and routing. It is possible to add central structures (arbiters) for the synchronization of the communication between agents. Such structures normally require the routing of protocol and/or data lines between the central structure and the agents. Since the arbiter is a central block, the routing of lines from the bus controllers to this central block may not be acceptable for some system implementations where the bus controllers are physically distant, since this increases routing area and usually require premium routing tracks.

978-1-7281-3147-4/20 $31.00 © 2020 IEEE

The total silicon area of any bus depends on the area of the bus controllers (A_D), the area occupied by shared structures in the system (A_S), and the routing area (A_R) which will depend on the floor-plan of the specific implementation. The total area (A_T) for a bus with M interfaces (or agents) can be described as

$$A_T = M \cdot (A_D) + A_S + A_R \quad (1)$$

In order to implement a bus architecture, an additional block to translate between the interfaces and protocols must be added to the bus controllers, so that in terms of area the translation between different bus implementations, it can be seen as a constant added to A_D.

There is an inverse relation between the number of lines shared by the bus controllers and the required implementation area A_D, since the use of centralized structures implies more shared signals. The use of distributed structures, in contrast, implies the duplication of logic structures in the bus controllers. In the case of the serial implementations, the reduction of the number of wires implies an increases of the area for the bus controllers. For the routing area A_R and the area of the shared structures A_S, an inverse relation with the number of lines required for routing is appreciated. In terms of power consumption, one can postulate that the power consumption increases with the area of the system and its throughput, assuming the maximum utilization of the throughput of the micro-architecture in a pessimistic test scenario, since this implies the maximum activity factors for all the nodes.

Every bus has a "minimum latency" (T_{min}), consisting of the minimum time required to transmit a flit of information. Assuming that every T_{min} it is possible to transmit n bits of flit, with k bits of payload and $n - k$ bits of overhead in the flit, it is possible to define the throughput of the bus (R) in bits/second as

$$R = \frac{k}{T_{min}} \quad (2)$$

Based on Eq. (2) and changing the minimum latency (T_{min}) to a minimum latency in clock cycles ($T_{C_{min}}$), where $T_{C_{min}} = T_{min}/T_{clk}$, it is possible to define the "Throughput per clock period" (R_C) in bits/T$_{clk}$ as

$$R_C = \frac{k}{T_{C_{min}}} \quad (3)$$

If there are L data/protocol transmission lines in a bus, a metric called "Efficiency per line per clock cycle" (η_{LC}) may be proposed to evaluate any bus implementation. This metric counts the number of useful bits transmitted by each shared transmission line of the bus per clock cycle, as given by

$$\eta_{LC} = \left(\frac{R_C}{L}\right) \quad (4)$$

Notice here that L also accounts for the lines communicating each bus controller to the shared structures. For an ideal digital bus, $\eta_{LC} = 1$, i.e., every shared line in the bus sends one bit of payload every clock cycle, and therefore there are no lines or clock cycles "wasted" in the synchronization of the flits, or dedicated to the transmission of headers for the bus protocol (assuming the bus is working at its maximum

transmission capacity). This metric considers inefficient implementations both in terms of the protocol and the micro-architecture. Using η_{LC}, it is therefore possible to evaluate, for instance, different micro-architectural implementations of the same standard in terms of their area of implementation before synthesis, assuming they work at the same clock frequency. Based on Eq. (1), one can postulate that:

$$
\begin{aligned}
A_D &\approx \alpha \cdot \eta_{LC} \\
A_S &\approx \beta \cdot (\eta_{LC})^{-1} \\
A_R &\approx \gamma \cdot (\eta_{LC})^{-1}
\end{aligned}
$$

where α, β, and γ are constants related to the micro-architecture and they must be estimated accordingly to a particular micro-architecture.

III. EVALUATION OF THE PROPOSED METRIC

A SystemVerilog interconnect library was developed in order to generate the buses to be evaluated. Once the RTL is generated from this custom library, it is embedded into a testbed, intended to reach the highest possible dynamic power consumption for the micro-architecture, while also reaching the maximum data transfer rate of the system. This generates a pessimistic power scenario. The results of the activity factors for every signal are recorded and feed into an RTL compiler and synthesizer tool, in order to generate a gate level netlist, from which accurate area and power estimations may be extracted.

Four micro-architectures were implemented, serial and parallel buses with and without central arbiters. In the serial bus without central arbiter (SB), one line transmits data and three are used for signaling; in the serial bus with central arbiter (SBA), two additional connections per bus agent to the arbiter are required. The parallel versions without arbiter (PB) has a wire per bit in the flit and additional bits for the synchronization of the data flow, whereas, the version with central arbiter (PBA) require additional n plus two bits per bus controller to coordinate with the central arbiter.

Table I compares T_{Cmin}, L and η_{LC} for each bus. In every bus in this work, k is equal to the number of bits in the flit minus eight corresponding to the identifier of the intended receiver of the message. In the SB implementation, the synchronization between agents is accomplished using a "token pass" strategy; a wire called "turn_change" is used to pass the token between agents in a round robin arbitration style, signaling when one agent finishes using the bus; an additional wire called "bus_busy" is shared between agents to inform when the bus is used and the data is transmitted serially using one wire. In order to transmit a single flit, four clock cycles are required for the synchronization of the data and an additional clock cycle s required for each one of the bits in the flit. Based on this, we can conclude that for SB, $T_{Cmin} = n + 4$ and $L = 3$.

For SBA, the transmission synchronization between agents is centralized in the arbiter, using "request" and "grant" signals; also, a single line is used for the transmission of data;

978-1-7281-3147-4/20 $31.00 © 2020 IEEE 14

TABLE I
PARAMETERS FOR THE CALCULATION OF THE "EFFICIENCY PER LINE PER CLOCK CYCLE" (η_{LC}).

Serial	SB	SBA
T_{Cmin}	$n+4$	$n+3$
L	3	$1+2\cdot M$
η_{LC}	$\frac{n-8}{3\cdot(n+4)}$	$\frac{n-8}{(1+2\cdot M)\cdot(n+3)}$
Parallel	**PB**	**PBA**
T_{Cmin}	5	5
L	$n+3$	$n+log_2(M)+M\cdot(n+1)$
η_{LC}	$\frac{n-8}{(n+3)\cdot 5}$	$\frac{n-8}{(n+2+log_2(M)+M\cdot(n+1))\cdot 5}$

3 cycles are required for the synchronization of the agents and an additional cycle is required for the transmission of each bit in the flit; therefore, for SBA assuming "M" agents connected to the bus, we can say that $T_{Cmin} = n+3$ and $L = 1+2\cdot M$.

For PB, the implementation is equivalent to the one described for SB but, instead of having one line for the serial data, it requires one line per bit in the flit to transmit in parallel; in this case, $T_{Cmin} = 5$, with four cycles for synchronization and one for the transmission of the flit, and $L = n + 3$. PBA is different from SBA because, instead of simply using "request" and "grant" signals in the arbiter, routing and flit transmission are centralized in the arbiter structure to simplify the controllers. Each bus controller share one line per bit in the flit (with a size of "n") for the incoming data and two additional control signals: "push", used to signal the receiver controller that the data can be saved and "pop", used to signal the driver controller that the data has been transmitted. Instead of passing the token using only one line, the token is signaled by a binary number of pointing to the ID of the controller which hold it; all the previously described signals are driven by the central arbiter. Additionally each controller must drive n lines to the arbiter with the output data plus one additional line signaling if there are pending flits to send. We have that $L = n + 2 + log_2(M) + M(n + 1)$ and the number of cycles required to send one flit is the same as in PB, so $TC_min = 5$. Finally, η_{LC} is estimated using Eq. (4).

Assuming a 1kHz clock and four agents in the system the transfer rates for the serial and the parallel buses are presented in Fig. 2. We use here 1KHz in order to get a result that may be easily extrapolated. Based on η_{LC}, one can estimate R for each one of the implementations using

$$R_c \frac{\eta_{LC}\cdot L}{T_{clk}} \tag{5}$$

from where it is easy to show that, in the best of cases, as the size of the flit increases, R_c is constrained for serial buses to the clock frequency of the clock. Meanwhile, for parallel buses, the transmission rate increases linearly with the number of bits in the flit, with a slope of $1/(R_C \cdot T_{CLK})$.

Figure 2 shows that the throughput for both serial buses is equal. Both parallel buses exhibit equal throughput as well. However Fig. 3 shows that η_{LC} is better for buses with a distributed arbitration. This because they require fewer routing between agents, while in the cases with central arbiter the additional lines required to communicate the central block with every agent in the system negatively impacts the metric. The proposed metric η_{LC} is limited to a finite value as the number of bits in the flit increase. These maximum values can be estimated as:

$$\lim_{n\to\infty} \eta_{LC}(n) \tag{6}$$

with η_{LC} converging to $1/(1 + 2 \cdot M)$ for the SBA bus implementation and to $1/3$ for the SB, as the number of bits in the flit increases; for the PBA bus, η_{LC} grows asymptotically to $1/(5 \cdot (1 + M))$, while for PB it grows asymptotically to $1/(5)$. Notice that, for buses with central arbiter, η_{LC} decays with the number of agents connected, because the number of routed lines increases while, for the distributed implementations, it converges to a constant because of the fixed number of shared lines.

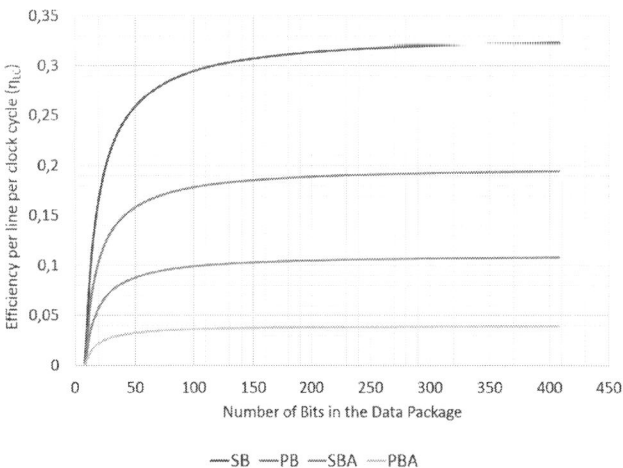

Fig. 3. Comparison of η_{LC} for each bus, assuming $M = 4$. Observe how η_{LC} converges to $1/(1 \cdot 2 \cdot M)$ for a SBA bus implementation and to $1/4$ for a SB, as the number of bits in the flit increase; in the cases of the parallel buses, η_{LC} for a PBA converges to $1/(5 \cdot (1 + M))$ and to $1/(5)$ for a PB.

Based on Eq. (1), the η_{LC} presented in Table I and the relations postulated for α, β and γ, it is possible to infer an

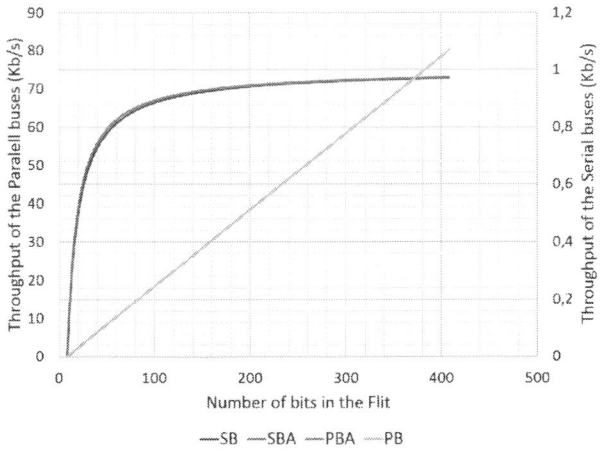

Fig. 2. PB, PBA, SB and SBA bandwidths assuming a 1kHz system clock with 4 agents connected to the bus.

equation for the area of each micro-architecture; notice that the routing area A_R depends on the floor-plan of the intended implementation, and its analysis is left as future work. Solving for $n = 32$ and taking into account only the cell placement area, the area for the different buses may be inferred as a function of the number of agents connected to the bus (M), such that

$$A_{SB} \approx \frac{2\alpha}{9} \cdot M + \frac{9\beta}{2} \tag{7}$$

$$A_{PB} \approx \frac{24 \cdot \alpha}{175} \cdot M + \frac{175 \cdot \beta}{24} \tag{8}$$

$$A_{SBA} \approx \frac{35 \cdot \beta}{12} \cdot M + \frac{12\alpha}{35} + \frac{35 \cdot \beta}{24} - \frac{12 \cdot \alpha}{70M + 35} \tag{9}$$

$$A_{PBA} \approx \frac{11\beta M}{11} + \frac{8\alpha}{11} + \frac{3\beta}{2} - \frac{96\alpha}{121M + 132} \tag{10}$$

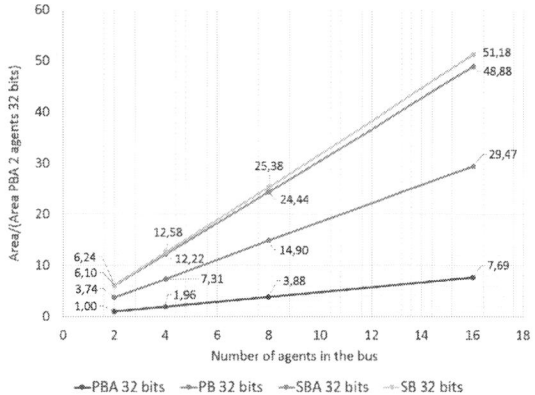

Fig. 4. Comparison of total cell area required for four 32-bit buses (@20MHz clock), using relaxed timing convergence constraints with a $0.18\mu m$ standard cell library. Area of the buses is normalized to the area of a PBA with two agents.

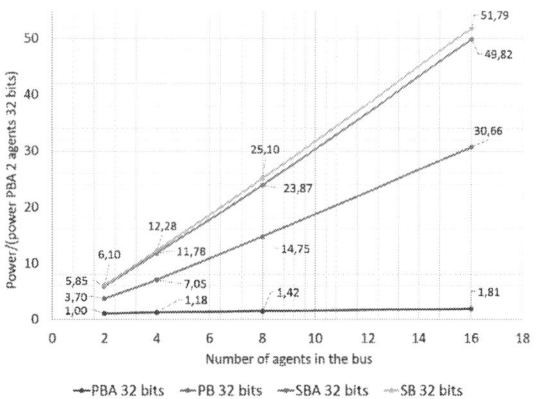

Fig. 5. Comparison of the dynamic power consumption for four buses (@20MHz clock), using relaxed timing convergence constraints with a $0.18\mu m$ standard cell library. Dynamic power consumption of the buses is normalized to the dynamic power of a PBA with two agents.

Depending on the design's fabrication technology, standard cells libraries, clock speed, placement and routing used, the area of the interfaces may change even using the same RTL description. In this work all the buses run at a 20MHz clock, using typical conditions for the standard cells library in power tests. One may assume that all the implementations scale similarly if any of the previously mentioned conditions vary, keeping the results presented in this work relevant. This statement requires nonetheless further exploration.

For the distributed implementations (SB and PB), the area equations predict a linear growth of the cell placement area, as the number of agents connected to the bus increases and, in the case of buses with central arbiter, an additional term that is inversely proportional to the number of agents is added to the linear equation. These terms converge to zero as the number of agents increase, so an approximate linear behavior is also expected. Figure 4 compares the total cell area required for different 32-bit buses, with their area normalized to that of the PBA with two agents. Notice how the predicted linear behavior matches the simulated one. Figure 5 compares power consumption. Here, leakage power is not significant and dynamic power is dominant. As expected, there is a close relationship between area and power consumption. The values of α and β for area and power approximations may be calculated using Eqs. (7-10) and a linear regression of the simulated data.

IV. Conclusions

A metric called "Efficiency per line per clock cycle" (η_{LC}) is presented as the first attempt to provide a semi-analytical approach for pre-silicon evaluation of digital on chip bus micro-architectures. Based on post synthesis simulations, we have found that the proposed metric in terms of area and power consumption offers an early estimate for designers when choosing a particular bus architecture, without the need to have a working RTL design. Further research is required to fully validate the metric using different floor-plans, synthesis algorithms, variations in the number of bits in the flit, and bus protocols.

References

[1] M. Mitić and M. Stojčev, "An overview of on-chip buses," *Facta universitatis-series: Electronics and Energetics*, vol. 19, no. 3, pp. 405–428, 2006.

[2] P. P. Pande, C. Grecu, M. Jones, A. Ivanov, and R. Saleh, "Evaluation of mp-soc interconnect architectures: a case study," in *4th IEEE International Workshop on System-on-Chip for Real-Time Applications*, July 2004, pp. 253–256.

[3] H. Li, X. Liu, W. He, J. Li, and W. Dou, "End-to-end delay analysis in wireless network coding: A network calculus-based approach," in *2011 31st International Conference on Distributed Computing Systems*, June 2011, pp. 47–56.

[4] M. Bakhouya, S. Suboh, J. Gaber, and T. El-Ghazawi, "Analytical modeling and evaluation of on-chip interconnects using network calculus," in *2009 3rd ACM/IEEE International Symposium on Networks-on-Chip*, May 2009, pp. 74–79.

[5] V. Zyuban and P. Strenski, "Unified methodology for resolving power-performance tradeoffs at the microarchitectural and circuit levels," in *Proceedings of the 2002 International Symposium on Low Power Electronics and Design*, ser. ISLPED '02. New York, NY, USA: ACM, 2002, pp. 166–171. [Online]. Available: http://doi.acm.org/10.1145/566408.566451

A compact functional verification flow for a RISC-V 32I based core

Roberto Molina-Robles*, Edgar Solera-Bolanos*, Ronny García-Ramírez*,
Alfonso Chacón-Rodríguez*, Alfredo Arnaud† and Renato Rimolo-Donadio*
*Escuela de Ingeniería Electrónica, Tecnológico de Costa Rica
†Depto. de Ingenieria Electrica, Universidad Catolica del Uruguay
{rmolina, esolera ,rgarcia, alchacon, rrimolo}@tec.ac.cr
aarnaud,@ucu.edu.uy

Abstract—The structure of a functional verification flow used for the design of a RISC-V core is presented. The paper offers a guide on the test-planning used and details of the flow architecture, showing how to integrate the Universal Verification Methodology with the required, reference models, while implementing key futures in standard verification environments, such as testing regressions and code and structural coverage. The designed flow is compact yet efficient, making it affordable for small design teams, without requiring extra investment other than the already necessary licenses for RTL synthesis and the eventual fabrication of the chip.

Index Terms—Functional Verification, RISC-V 32I, UVM, SystemVerilog, EDA tools, architecture, test generation, processor, compiler, simulation, coverage, regression, reference model.

I. INTRODUCTION

Functional verification is hardly a new topic. There is a solid standard [1], and plenty of teaching resources available (see for instance [2] for a comprehensive site with plenty of free resources). And there exists some recent literature with examples that can provide guidance to non-experienced verification teams. Some works, for instance, show how to implement verification environments with or without UVM, such as [3], where a custom environment was proposed to improve coverage, or [4] and [5], where UVM was applied to different RTL blocks. Yet, most of the documentation available points towards the use of massive, highly integrated frameworks attached to a particular methodology—typically derived from the Universal Verification Methodology (UVM)—, where the access to commercial tools and personnel is not a limitation, and the departure specifications for the expected results are readily defined (or at least, do not depend directly on the verification team itself). Yet, small design teams usually do not have the budget to tackle the verification problem using such an approach, particularly when the design and verification process must be handled by the same people. This means finding ways to optimize hardware and software resources, through the use of open source tools whenever possible, while providing with a flexible environment that can easily be migrated from a project to another. Now, some examples of small teams using functional verification for their chips approach can be found. Yet, to our knowledge, most of the goal specifications in those papers were already defined by the use

of a standard given architecture (SiFive's Rocket-chip), written in Chisel. This was not the case here, being Siwa an architecture written from scratch, with several modifications from the RISC-V 32I standard mandated by the ultra low power specifications of its intended application. This paper's main purpose is to document the implementation of a functional verification environment used for the pre-silicon verification of a RISC-V 32I based processor, called Siwa, developed by a small team of only six people as the main controller core of a medical implantable tissue stimulator (see [6] for details on the processor and the medical device system itself). The present work gathers strategies as those used in the previous references, and incorporates them not only for the verification of architectural blocks, but also, adds up the use of reference models, the integration of custom architectures and the automation of regressions for random verification, topics often missing in the literature.

As such, the structure here presented allows for functional verification at different hierarchical levels, using oriented and random-constrained tests, based whether on custom reference models or requiring the incorporation of code simulators serving as golden vectors generators, while using regression systems for extending code coverage. All these in an environment compact and versatile enough for a verification team of only three people.

This paper is organized as follows: Section II describes the tools and resources selected for this work. Section III details the establishment of the verification plan. Section IV describes the proposed verification flow. Section V presents the results. Lastly, Section VI highlights the main conclusions of the work and discusses some future work.

II. DEFINITION OF TOOLS AND RESOURCES

Small IC design teams with modest financial resources typically require affordable yet competitive, low maintenance tools. Yet, one cannot always resort to the open source community for such tools, although there is a strong movement pushing in that direction of open hardware design, as for instance the Linux Foundation CHIPS Alliance. There is also a lack of technology kits for open design tools, as foundries base their production environments on the three major EDA providers: Mentor Graphics, Synopsys and Cadence. Particularly, our group has a Synopsys license available. In the case here presented, selecting SystemVerilog (SV) as specification language, and UVM as the verification methodology was straightforward. SystemC was discarded due to its lack of support from the Synopsys synthesis tools and the libraries flow provided by the foundry for the project, with the added extra of SV already having a UVM library incorporated, and being supported by most of compilers/simulators. This meant using Synopsys VCS for the compilation, simulation and coverage processes and Synopsys DVE as the Waveform Viewer and Coverage Analyzer of choice. This mainly because of the tools' easy interfacing with the Design Compiler and IC Compiler flows. Alternative simulators (such as Verilator) and wave analysis tool (GTKwave), nonetheless may be used.

Concerning the setup of the regression platform, Linux's Cron utility and basic Bash scripting were selected, even though there are other options such as Jenkins and Bamboo that are typically more favored by industry because of their wide array of features and support for large development teams. Cron, nonetheless, is easier to setup and less demanding in terms of computation resources.

III. First Steps: Verification from Scratch

Before addressing the verification architecture, a road-map is first created. This means studying the design via specification documents and industry standards, and coordinating with the core's architects and designers. Having adequate knowledge about the functionality of the chip to be verified accelerates the verification process. Coding can be tackled once the verification plan is ready. Methodology, verification architecture, test plans, resources and tools, time-lined efforts, coverage points and others aspects are expected sections of the verification plan, and should be reviewed several times with architects, designers and other verification engineers.

Regarding this work's implementation, the environments were custom built for a processor based on RISC-V 32I standard [7], focused on medical applications [6], and most of them developed under the UVM standard.

IV. The Verification Architecture

Since the processor was small, verification efforts were into only two hierarchical levels. The lower level for block verification and the higher level for chip verification. A simulation-based verification flow must include several characteristics, which can be found here [8].

A. Block-Level Verification

The selected blocks submitted for verification were: an Arithmetic-Logic Unit (ALU), a Memory-Bus Controller (MBC), a System Bus and a Universal Asynchronous Receiver-Transmitter port (UART). The ALU and the UART blocks were designed using a standard register-based RTL approach, with minor custom modifications. The MBC and the bus used a latch-based micro-architecture, for area and power reduction. Either way, for block-level verification then,

UVM was used to implement the verification architectures. The remaining block was an SPI module that was stimulated at chip-level, however, its complete verification will be done in further spins. A Black-Box philosophy was selected for the block-level verification efforts. In [9], there is a clear explanation of the advantages of this type of verification. Figure 1 shows the generalized block diagram for the block-level architecture, based on [1].

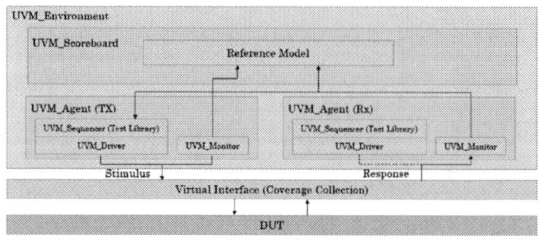

Figure 1. A testbench implemented using a UVM architecture for block-level verification. The quantity of agents varies depending on the modularity of the design block interfaces. The sequencer calls sequences built with sequence items to form a test. The desired test is called from the command line.

There were small variations in the architecture's implementation for each individual block, but the general idea remains the same. For example, the TX Agent and RX Agent from Fig. 1 were here fused into a single agent for the ALU and the UART, since those blocks had few interfaces and their functionality were simple enough. In contrast, the MBC and the Bus had more than two agents because of their several interfaces. This type of modifications are supported inside the UVM standard, but one must remember that the less agents one has, the easier its environment construction but the harder its maintenance. The ALU and the UART are blocks with standard interfacing, not prone to change in further micro-architectures, and thus, agents may be considered stable as well. Meanwhile, the MBC and Bus are blocks that may change depending on features that might be added in further spins. If one separates the interfaces in features/devices and connect an individual agent to it, then one can only adjust the associated agent instead of modifying a more complex agent connected to all interfaces at once.

At the block-level, the reference model used was transaction-based [9]. That is, that at every transaction between the verification environment and the DUT, a checking is being made between the custom reference model's prediction and the actual results of the DUT. This reference model was implemented in SV inside the scoreboard, based on a written specification to avoid similarities with the RTL implementation as much as possible.

B. Chip-Level Verification

Chip-level verification means that the DUT is the complete RISC-V core. Here, a first consideration is that the core runs programs that cannot be totally randomized: for instance, memory addressing instructions have limited valid ranges, and certain registers have pre-defined functions. Second, the core has concurrent interfaces (UART, SPI and GPIO). Third, the easiest way to debug a processor is looking into the core's register bank. However, none of these registers can be externally accessed, except indirectly via the UART or SPI ports. Finally, the moment when the checking with the reference model occurs must be selected carefully according to the micro-architecture; that is, that the time needed to execute instructions varies depending on the core structure (simple scalar multicycle, pipelined scalar, pipelined superscalar, etc.).

These restrictions imposed a Grey-Box approach as the chip-level verification philosophy, where the register bank is accessed via a backdoor. A "Golden Reference" methodology was selected, as recommended by [9], although it is possible to use a transaction-based reference model if the checking occurs at the end of each instruction cycle. Test program generation was handmade at this level. Figure 2 shows the custom architecture of the chip-level verification environment.

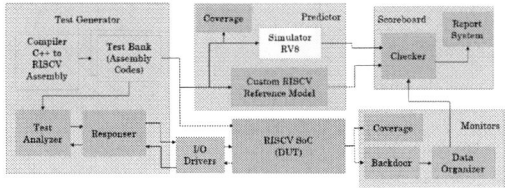

Figure 2. Custom architecture implemented for pre-silicon testing of a RISC-V core. The test generator calls a program created with the compiler and loads it inside the DUT. The responser communicates with the DUT if it is instructed in the program. Monitors are connected via backdoors to registers and specific control signals. Blue blocks and the DUT were written in SystemVerilog.

A program is loaded and executed in Siwa's RTL model. Information from the data and control-status registers (CSRs) information is stored in an array at the end of each instruction cycle. Simultaneously, a reference model predicts the correct result that ought to be stored in each register for every instruction, and stores it as well in another array. The final data arrays are compared after the testing program ends.

C. Generation of the Reference Model for Chip-Level Verification

Contrary to the block-level reference models, custom built using a specification document, a standardized reference model was used for chip-level verification. In order to build such custom reference model, the logic described in Fig. 3 is followed. The RISC-V simulator RV8 [10] is used in tandem with a custom reference model written in SV, validated itself against RV8 simulation results. The goal was to construct a model capable of predicting results stored into the register bank for each instruction. The custom reference model is used for Siwa's custom operations that do not follow the RISC-V standard; specifically: a smaller set of CSR and a restricted memory map (8kB).

D. Test Generation

The test generation was handled differently depending on the hierarchical level. All tests for block-level DUTs were randomly generated and constrained. The verification plan

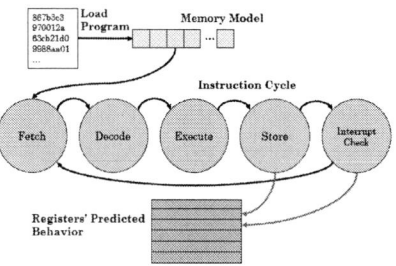

Figure 3. RISC-V reference model flow diagram. This model is called before loading the Flash memory. The predictor also helps to determine when a test should end, if the RTL does not reach this point, it means that something went wrong. Each of these steps conforms a typical instruction cycle.

previously designed specifies the necessary tests. For instance, a test for each arithmetic-logic functionality of ALU was implemented, for each type of transactions through the UART at different speeds and configurations. Read and write tests for the MBC and tests that emulate data flow traffic through the Bus were also created.

At chip-level, however, random instruction generation was more complicated. Since instructions must follow the ISA standard and the program needs a coherent intention to be able to run an application, the tests were completely oriented, hence, the random factor was taken out. Test selection comes from the verification plan. This meant tests were implemented for each instruction in the RISC-V 32I ISA [7], for each port and its intended functionality, for each type of core interrupt, and tests that increase coverage over the register bank and memory space were also needed. Each of these tests were written as individual programs to be loaded and run into the core. These tests were handmade and developed using a C Assembler compiler, from the Sifive Toolchain [11], [12]. Fig. 4 shows a diagram the tests order of execution, as they feed the verification environment.

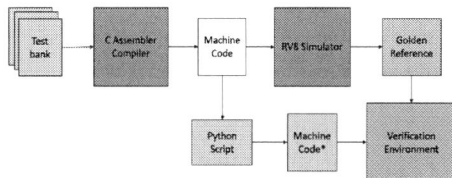

Figure 4. Diagram that depicts how tests are handled prior the beginning of the simulation. Machine code marked with a * has been edited with a Python script, in order to match the Flash memory model requirements.

The resulting text file with the program was modified afterwards via script to adjust it, so the SV model of a Flash memory could read it. Then, the core boots from that Flash memory connected to a SPI (Serial Peripheral Interface) port. Finally, the rest of the simulation and the checking occurs as explained on the previous sections.

E. Regression System

Regressions in functional verification are similar to test farms, where multiple tests are run one after another. Some are pseudo-random, and therefore, are linked to a seed. Others, are oriented tests running a particular important feature. The proposed flow integrated a regression platform, depicted in Fig. 5.

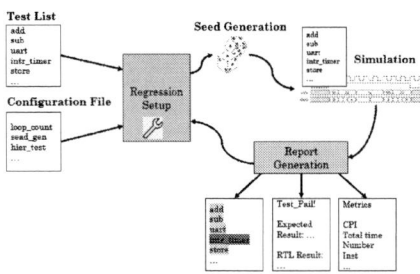

Figure 5. General structure of the regression system. Each test is run according to a list of tasks, with a configuration file defining specific aspects of the regression, such as loop quantity and type of coverage collected.

Basic shell scripting was used for coding the regression platform, following the general structure shown in Fig. 5. Seeds are generation using Python. For oriented tests at chip-level, the RISC-V compiler is invoked for test generation with RV8 providing the golden reference construction, prior to invoking VCS for simulation and coverage collection. A report is created or updated with the resulting status for each test. Once a batch of tests is finished, a new batch is prepared and executed using a different seed. The required number of iterations is specified by the user, according to the goal in terms of coverage.

V. RESULTS

The proposed verification structure is fully functional and can be replicated to equivalent designs. More than 100 tests were implemented in total, some of them with multiple seeds, for at least 5 different DUTs with independent functional and structural coverage metrics.

An example of the regression console is given in Fig. 6. Several reports are generated by the regression, including information such as: test status with its respective seed, performance parameters such as execution time or Clock Per Cycle (CPI), and comparisons between the predicted and the RTL register bank for each individual test.

Figure 7 shows an example of coverage results, extracted from the Synopsys DVE coverage tool.

VI. CONCLUSIONS

A compact, affordable functional verification flow has been generated and used for the verification of a small RISC-V 32I based micro-controller, as a base case. The flow follows a functional verification strategy the includes the development of a test methodology and verification architecture, and is capable of carrying out hierarchical verification, reporting coverage metrics and performing intensive, randomized regressions.

Figure 6. Screenshot of the regression console. Several reports are generated, and the execution is scheduled via Linux's Cron.

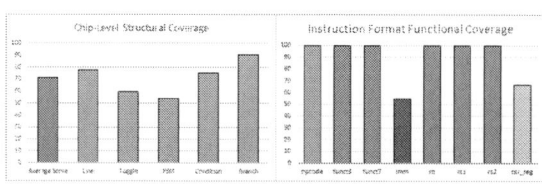

Figure 7. Accumulative coverage results obtained with the custom verification environment. The graph to the right shows the functional coverage of the RISC-V instruction formats [7]. The graph to the left depicts the average structural coverage obtained for chip level verification.

The flow can be extended to other digital designs and can incorporate alternative tools if required. Future works include adding the option for formal verification tests to the flow, and the possibility of generating random RISC-V coherent programs.

REFERENCES

[1] *Universal Verification Methodology (UVM) 1.2 User's Guide*, Accellera, 2015.

[2] Doulos. (2019) Uvm knowhow. [Online]. Available: https://www.doulos.com/knowhow/sysverilog/uvm/

[3] R. Yang, L. Wu, J. Guo, and B. Liu, "The research and implement of an advanced function coverage based verification environment," in *2007 7th International Conference on ASIC*, Oct 2007, pp. 1253–1256.

[4] V. B and B. Bala Tripura Sundari, "Uvm based testbench architecture for coverage driven functional verification of spi protocol," in *2018 International Conference on Advances in Computing, Communications and Informatics (ICACCI)*, Sep. 2018, pp. 307–310.

[5] T. M. Pavithran and R. Bhakthavatchalu, "Uvm based testbench architecture for logic sub-system verification," in *2017 International Conference on Technological Advancements in Power and Energy (TAP Energy)*, Dec 2017, pp. 1–5.

[6] R. García, A. Chacón, R. Castro, A. Arnaud, M. Miguez, J. Gak, R. Molina, G. Madrigal, M. Oviedo, E. Solera, D. Salazar, D. Sánchez, M. Fonseca, J. Arrieta, and R. Rimolo, "Siwa: a RISC-V platform in a $0.18\mu m$ HV CMOS process for implantable medical devices," submitted.

[7] A. Waterman and K. Asanovic, *The RISC-V Instruction Set Manual Volume I: User-Level ISA*, SiFive Inc., CS Division, EECS Department, University of California, Berkeley, 5 2017, an optional note.

[8] "Ieee standard for the functional verification language e," *IEEE Std 1647-2016 (Revision of IEEE Std 1647-2011)*, pp. 1–558, Jan 2017.

[9] B. Wile, J. Goss, and W. Roesner, *Comprehensive Functional Verification: The Complete Industry Cycle (Systems on Silicon)*. San Francisco, CA, USA: Morgan Kaufmann Publishers Inc., 2005.

[10] Risc-v simulator for x86-64. [Online]. Available: https://rv8.io/

[11] Risc-v gnu toolchain. [Online]. Available: https://github.com/sifive/riscv-gnu-toolchain

[12] Risc-v elf to hex converter. [Online]. Available: https://github.com/sifive/elf2hex

Author Index

Alfaro-Badilla, Kaleb	PRIME-1
Arnaud, Alfredo	PRIME-5
Bampi, Sergio	PRIME-3
Chacon-Rodriguez, Alfonso	PRIME-1, PRIME-4, PRIME-5
Chiossi, Maximiliano	PRIME-2
García-Ramírez, Ronny	PRIME-1, PRIME-4, PRIME-5
González-Gómez, Jefferson	PRIME-1
Klimach, Hamilton	PRIME-3
Miguez, Matías	PRIME-2
Molina-Robles, Roberto	PRIME-5
Rimolo-Donadio, Renato	PRIME-1, PRIME-4, PRIME-5
Salazar-García, Carlos	PRIME-1
Solera-Bolaños, Edgar	PRIME-5
Strydis, Christos	PRIME-1, PRIME-4
Teles Medeiros, William	PRIME-3

IEEE
445 Hoes Lane
Piscataway, NJ 08854-4141

ISBN 978-1-7281-3147-4

9 781728 131474